The Patrick Moore Practical Astron~

For further volumes:
http://www.springer.com/series/3192

From Casual Stargazer to Amateur Astronomer

How to Advance to the Next Level

Dave Eagle

 Springer

Dave Eagle
Higham Ferrers, UK

ISSN 1431-9756
ISBN 978-1-4614-8765-4   ISBN 978-1-4614-8766-1 (eBook)
DOI 10.1007/978-1-4614-8766-1
Springer New York Heidelberg Dordrecht London

Library of Congress Control Number: 2013947792

Springer is part of Springer Science+Business Media (www.springer.com)

Preface

Astronomy is one of the few sciences where the amateur can still make a valuable contribution to real science. Today there are a growing number of professional-amateur collaborations where professional astronomers rely on amateurs to collect and process data. By observing the sky themselves, amateurs can still discover or make an extremely useful contribution to making new discoveries.

Like any relationship, successful astronomical observing takes some level of learning, commitment, communication and investment in both time and effort. Only by doing this will you get the very best out of our fabulous hobby and that intimate partnership to mature properly.

As a person who is interested in astronomy, do you feel that you could get involved yourself? If not, why not? Is this because you don't have enough knowledge? Or is it because you don't feel you just do not have enough experience?

There is really only one way to get the knowledge and experience needed, and that's to get out there and keep observing yourself.

Much as the term "stargazer" is much loved around the world, the term "amateur astronomer" does sound far more professional, but at the same time intimidating.

Within these pages you will take a long hard look at where you currently lie along your path of discovery. It will clear the way along this path, helping to steer your journey with the least pain, or at least help you to avoid some of the major pitfalls.

You will also find lots of hints and tips as well as sources of information so that your knowledge and observing confidence matures and flourishes.

By the end of the book, you really will feel that you too consider yourself an amateur astronomer, rather than a stargazer.

Higham Ferrers, UK

Dave Eagle

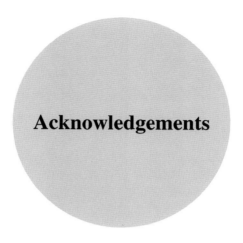

Acknowledgements

Not only is this book the result of years of experience and frustration from years of observing from under the cloudy skies of the UK, it is also the result of the input of many people over many years of observing. So I really cannot take all the credit for it.

It would not have been possible without the assistance and input of quite a number of people. This ranges from the many people I have met over the years at local astronomy clubs, particularly Bedford Astronomical Society. Many of these have become very good friends. Along the way they taught me a thing or two about astronomy as I journeyed my own path of astronomical discovery.

I would also like to thank Dusko Novakovik and Nick Hewitt from Northampton Natural History Society, Astronomy Section. Diligently reading my ramblings, they picked up quite a few of my gaffs and spelling mistakes, gently advising me to make a few changes. I hoped this would prevent me from making myself a laughing stock. Let's hope it has worked.

I cannot praise or thank my illustrators enough.

Seb Jay, for his major contribution on the chapter about drawing in this book and his detailed astronomical drawings, which really do put most of mine to shame.

The wonderful line drawings of me by the telescope were produced by Gita Parekh. I think they really stand out and give the book that little personal touch.

Most of all I would like to thank my loving wife, Sue. She endures the passion I have for my favourite hobby, (one of far too many), as well as the dome taking up a good proportion of our back yard. Sue has put up with a lot since we met, but throughout it all, she has always been extremely supportive in all my endeavours.

Now that this book is finally finished, Sue will be extremely relieved to have her husband back in the land of the living once more, instead of being a slave to his computer.

Higham Ferrers, UK   Dave Eagle

About the Author

Dave Eagle has been interested in astronomy for most of his life. As a boy, he followed with great excitement the landing of the Apollo missions, which sparked his enthusiasm even more. This set him well on the way to learning much more about the night sky as he started to explore it for himself.

In his mid-20s, Eagle established the Bedford Astronomical Society, holding the position of Secretary for many years before becoming Chairman. He also held the post of Handbook Editor for the Federation of Astronomical Societies for 3 years and is a Fellow of The Royal Astronomical Society. He is an enthusiastic ambassador for the subject, frequently giving talks to local astronomical societies, social clubs, schools, youth groups and other interested parties. He also encourages others to get out and observe, producing a monthly sky guide which is free to download from his website.

Eagle is a trained biological scientist, but after trying his hand at science teaching he eventually moved out of the labs and into the field of IT. He is fortunate enough to have his own small personal observatory in his back yard and considers himself a good all-rounder, enjoying all aspects of astronomy. Despite suffering from a reasonable amount of light pollution in his small town, he is still able to actively observe and image the sky.

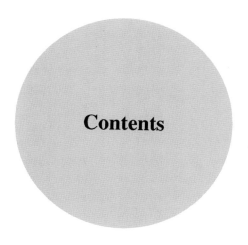

Contents

Part I

Developing Your Skills

Chapter 1

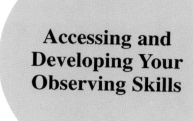

Accessing and Developing Your Observing Skills

1.1 Stargazer or Amateur Astronomer?

The word amateur stems from the French word Amour, meaning "Lover Of". And there is a whole army of amateur astronomers around the world who just love doing astronomy. They don't get paid for the privilege of experiencing the sky in all its glory, but by making detailed observations they do make a very important contribution towards the Science. These observations are especially useful when organized as a collective effort. Citizen science has really taken off in the last few years and the GAIA project will soon be producing so much data, that the professionals just will not have enough manpower to tackle all the data. They will rely on amateurs sitting on their computers at home. But it is under a dark sky that astronomy really comes alive. The fact that you have picked up this book, must mean that you are interested in taking the hobby a step forward.

We will assume that you have now had the astronomy bug for a while. Since that time, you will have found yourself passing through several stages in your chosen observing "career". You started casually looking at the sky and as your interest developed, you got a bit more intrigued. You started to read a lot more to feed your growing thirst for knowledge. Maybe you subscribed to astronomy magazines, drooling over pictures of beautiful objects. You then browsed longingly at a bewildering array of wonderful instruments designed, not only to reveal the universe to you, but seemingly to drain your bank account into the bargain.

How can there be so many telescopes and accessories to choose from, and of so many different types? After searching around for more information on the Internet

D. Eagle, *From Casual Stargazer to Amateur Astronomer: How to Advance to the Next Level*, The Patrick Moore Practical Astronomy Series, DOI 10.1007/978-1-4614-8766-1_1, © Springer Science+Business Media New York 2014

Fig. 1.1 How do you know where to look? (Courtesy of Gita Parekh)

you may have found yourself on one of the many astronomy forums. After asking many questions and doing even more research, you eventually plunged into purchasing that longed for telescope within your budget.

Since it arrived, it seemed that a complimentary shipment of clouds was also included with the deal, blotting out the night sky for what seemed like ages. Your shiny first telescope gathering dust in the corner of the living room, waiting for that vital clear night. Eventually the skies did clear enough to enable you to go out and obtain first-light with it. Excitedly you observed rugged features on the Moon's surface.

You gasped in awe as Saturn and its beautiful rings drifted into your field of view as well as some of the details on Jupiter's disk. It took you a while, but you found quite a number of Messier's fine deep-sky objects, some more easily than others. You may have even been lucky enough to have observed a comet or a faint supernova.

But now you've seen all this, where in heavens do you go from there?

Fig. 1.2 Image of the Moon just after 1st Quarter (Courtesy of the author)

Anyone who quickly expects to know all there is to know and see and do everything, in a very short space of time, is highly likely to find things are maybe just that little bit harder than first anticipated. Couple this with extremely high expectations, many potentially proficient observers quickly become disappointed and demoralized. Sadly in many cases, what at first started out as an exciting hobby, can quickly become tiresome, too much like hard work and decidedly boring. You can guarantee that the weather will also conspire against you, adding to the growing list of annoyances. All these will reduce the amount of time you are able or willing to spend pursuing the hobby and will dampen your enthusiasm. The telescope once so excitedly received starts to sit neglected once again gathering dust in the corner. After a while your previously treasured telescope soon finds itself on the second-hand market with your interest well and truly snuffed out. Your equipment will be well received by another newcomer to the hobby who, in all likelihood, may find themselves going round a similar cycle.

This book is aimed to help those who find themselves in this very position. This book will act as your personal observing buddy, helping you to re-discover the

Fig. 1.3 DSLR image of the galaxy pair M81 and M82 (Courtesy of the author)

excitement that got you interested in the first place. It will also lead you to sources of useful information and pearls of advice accrued over many years of observing experience. This will enable you to identify what it is you really enjoy doing and get the most out of your observing. After all, not everyone sees the appeal in tracking down faint fuzzy blobs, or separating close double stars. Others find that the Moon is a really bright nuisance, swamping out their real passion rather than a wonderfully detailed object in itself to behold.

This book aims to help start you on the road to confronting your perceptions offering you up new challenges for you to try. Helpful hints and tips have been included and there are a number of exercises to help you develop your observing skills and your overall awareness of the sky. In turn this should help you to develop your observing to a higher level, committing to the next. As you start to concentrate your efforts in a more planned approach to observing it will build your confidence, so that the hobby become much more rewarding and fulfilling.

Look at your attitude towards the hobby as it stands now and contemplate the following statements:

"Although it's clear out there tonight, it's freezing cold. Much better to stay in the warm in here watching TV",
"If I set everything up it will soon cloud over, so I just won't bother",
"It's always clear when I'm doing something else",
"That bright Moon is always interfering with my dark skies".
"It's far too much hassle to go out and set everything up".
"Those faint deep sky objects don't look anything like the Hubble Space Telescope images",

"Once you've seen one fuzzy blob, you've seen them all"
"Surely now I must have seen everything there is to see?".

If some of those statements sound far too close to home, and you find yourself saying at least one of them far too often, you have definitely reached the end of the honeymoon stage of the hobby. Excuses not to do observing come far too easily.

1.2 Perceptions and Expectations

The author was in discussion with a fellow observer one spring who stated: "There's nothing around at this time of year worth observing!". In the northern hemisphere, spring brings us the best of the distant galaxies in the evening skies with Leo, Virgo and Coma Berenices beautifully displayed. The lighter evenings may be starting to draw back in at that time of year in the northern hemisphere, but there is plenty to see if you take time to find out what is visible and actually get out there and observe. Needless to say, he left the hobby less than a year later and was never heard from again.

The biggest impression a beginning observer is often faced with about astronomy, especially when looking through astronomy magazines, is that to appreciate and enjoy astronomy properly, you have to spend a lot of money on expensive equipment. This is not necessarily the case.

This book will show you how you can gather information and step up your observing. This will enable you to get the most out of your existing equipment. You can just make a few small changes to increase your observing successes without having to spend a fortune.

After reading this book, you will have a greater understanding of the sky as well as understanding your observing interests and strengths. You will have developed better observing skills, regularly working from a personal observing list and will have the knowledge needed to get out and observe certain objects or phenomena giving you the best chance for a successful observation. This will in turn develop your confidence and you will soon start to observe objects you currently think are virtually impossible.

Anything you pursue will be always be dictated by your specific interests. So, before we do get out and observe, let us investigate what skills you already have and find out what your real observing interests are.

1.3 Assessing Your Skills and Identifying Your Interests

In this chapter we will identify the skills you already have, looking at what really interests you and identifying what may be stopping you from getting the most out of your hobby.

Unfortunately this chapter is all about making and using lists, but please persevere as it should help you determine what you enjoy and should start to make your observing much more productive.

How do you feel about the hobby?

As stated previously, astronomy is one of the few disciplines of science where the amateur can make a real contribution to our general understanding within the field. Professional astronomers are highly polarized, looking at specific objects and tightly bound within strictly regimented observing schedules, fighting for precious time on expensive instruments. The amateur on the other hand has the whole observable universe at his fingertips, pointing his telescope at a whim at whatever object takes their fancy at the time. This increases the chance of an amateur spotting something out of the ordinary and discovering something new. Being out there and observing will increase the chance of you making that new and exciting discovery. So why couldn't that be you? What is stopping you?

Exercise 1A: List Your Experiences of Astronomy

Start by sitting down and thinking about what first drew you into astronomy.

Make a list as prompted by the questions below. You could ask many more additional questions yourself.

How was your interest in astronomy first stimulated?
What exactly was it that appealed to you about the hobby at that time?
Why did you decide to start observing yourself?
What drove you to buy your first telescope?
What were you hoping to see or achieve?
Have you been pleasantly surprised at how much you have seen?
Do you frequently find yourself feeling more and more disappointed at what you cannot see?

Cast your mind back to the thrill you felt when you had your first views through a telescope. What went through your mind then?

How excited did you feel when something you were hunting down for ages suddenly appeared in your field of view?

What gave you the biggest buzz? Was it the intricate detail on the Moon and planets or tracking down and finding that elusive smudge of a comet?

Do faint fuzzy blobs like nebulae and distant galaxies really appeal?

What do you feel seems to cause you the most frustration when you are observing?

Has the way you feel about observing changed from when you first started out? How has it changed?

Does this change in attitude encourage or discourage you from observing?

If your feelings have changed, can you put your finger on what might have happened to cause this?

Has this change in attitude resulted in you doing more, or less observing?

If you do observe less, have you any idea what might be dampening your spirit? Is there a common problem that usually stops you seeing what you want? Have you identified anything that frequently puts you off going out observing?

Long term involvement in astronomy isn't for everybody. Let's face it, some people do have fleeting fancies, but as you have made the effort to pick up this book, it appears that you would like to stick with it.

Now you should have a good idea of how you currently feel about our hobby. Let us start by looking at what you have already achieved since starting out. You will probably be very surprised at how much you have already managed to observe and how much you have already developed.

Exercise 1B: What Have You Achieved Already?

Make yourself a list of some of the objects you have already seen.

Below are listed a few suggested questions that might help in your thinking. Again, there must be many other questions you could ask.

How many constellation patterns do you recognize? Can you name the constellations that are visible during the different seasons of the year? Have you observed any asterisms? What is the faintest star visible to the naked eye from your location? Have you ever estimated and followed the brightness of a variable star? Have you plotted the position of Barnard's Star on a star map and compared it to a previous observation? How many double or multiple stars have you resolved? How close a double star can you resolve with your current setup? Have you ever used a safe solar filter to observe sunspots and faculae on the Suns surface?

How many of the Moon's notable surface features have you identified? How many of the Moon's many phases have you actively looked at? Have you ever tried spotting the thinnest crescent Moon appearing in the evening sky, or disappearing in the morning? When was the last time you saw an occultation of a star, planet or an eclipse of the Moon? Have you identified the areas where the Apollo missions landed? Have you seen the mountain ranges around Mare Orientale?

How many of the other seven major planets have you observed? How much detail have you managed to see on them, or in the rings of Saturn? How many phases of Venus or Mercury have you observed?

Fig. 1.4 Webcam montage of the phases of Venus (Courtesy of the author)

How many satellites of Mars, Jupiter, Saturn, Uranus or even Neptune have you observed? Have you managed to observe Venus in daylight? Have you drawn the observed track of a planet or asteroids position on a map of the sky?

How many minor planets or asteroids have you identified? Have you ever estimated the brightness of an asteroid, or plotted the movement of one on a star map? Have you ever spotted a near-earth object?

How many meteor showers have you observed? Have you ever counted the amount of meteors seen in a night and plotted them on an all-sky star map? Have you observed a comet? If so, when was the last comet you observed? Have you drawn the track of a comet on the sky from your observations? Do you know how to estimate the brightness of a comet?

Have you seen a pass of the International Space Station? How about an Iridium Flare? Do you know when the International Space Station is going to pass across the Moon or Sun, or how to find out when it will?

Have you observed any other satellites or space probes passing by? Have you observed all Messier's 110 objects, or all 109 Caldwell's? How about Herschel's 2,511 Objects? Do all 7,480 NGC objects intimidate you? Have you seen the central star in the Ring Nebula, spotted a globular cluster around the Andromeda galaxy, seen nebulosity amongst the stars of the Pleiades star cluster or tried to view faint members of a distant galaxy cluster? Have you ever observed a supernova in a far off galaxy?

As you can see from the suggested list above, there is such a wide variety and a huge number of objects available to get out and observe. There are probably many others, which the author may have been forgotten from this list. With such a diverse range of objects, how can you ever really get bored at what you can see? There is always something else to observe that you have never looked at before.

We are meant to be doing this hobby because we enjoy it. Maybe you do not go out because of the weather. Maybe it is the wrong time of the year. Perhaps you are not interested in that aspect of astronomy at all or there is far too much going on in your day-to-day life (and we all have those sorts of times). Perhaps you have the impression that something will be too hard to observe or much too difficult to

Fig. 1.5 Webcam images of the International Space Station (Courtesy of the author)

achieve with your limited equipment or ability. Perhaps light pollution in your area prevents you from setting up and observing.

All of these situations sound far too familiar to most of us. We all have them no matter how advanced into the hobby we may be. There are also those nights when everything seems to plot against us, especially the weather, or technical hitches when nothing seems to work. These all seem to conspire against us from seeing what we set out to observe. Whatever it is, sit down and try to identify the root cause that is really stopping you from getting the most from the hobby. From the exercise above, have you singled out already one thing that may be preventing you from developing your interest further?

You should now have quite a comprehensive list. Now it is time to have a long hard look at your list now to sum up what you have found.

1.4 Taking Stock

How many things have you achieved since you started observing? Quite a few? However, how many on the authors suggested list above have you observed? Very few of them one would expect. This is written with much confidence because even a seasoned observer with after over 40 years of experience just could not have observed everything. It would not be from a lack of enthusiasm on the observer's part, but there is just far too much to observe out there to observe, and sadly not as many opportunities to enjoy them as one would like.

We are constantly bombarded from all directions with colorful Hubble images, showing our wonderful universe in all its glory. When you do look down the eyepiece of your telescope, you will never see anything approaching that view. We can sit on the Internet and see all these fantastic images. So why do we go out and observe ourselves?

A lot of the sense of achievement you feel is down to your expectations. Keep them too low and things soon become rather boring. Raise them far too high and expect to see Hubble-type views through the telescope, as much of the media tends to suggest, and you really will be highly disappointed with the experience.

Many observers are able to carry out their hobby despite problems like light pollution or only being able to afford relatively cheap instruments and accessories. So why do they seem to be so successful?

In many cases, it is sheer determination that those issues are not going to spoil their enjoyment of the hobby. They make the most of their particular situation and persevere where others give up far too quickly.

1.5 Some General Observing Rules

Before we go further, let us lay down some general observing comments.

- All celestial objects look brighter and much clearer the higher they are in the sky.
- Celestial objects are at their highest when they transit across the meridian.

- Hazy nights sometimes have steadier seeing and are frequently better for lunar and planetary observing.
- Clear, dark skies are better for faint deep-sky objects.
- Light pollution, haze or moonlight will reduce the number of faint objects you can see.
- Lighter nights during the summer months really restrict observing at temperate latitudes as the sky never gets properly dark. Twilight effectively lasts all night.
- Clouds will (usually) stop you seeing rare events, or appear as soon as you get everything set up ready.
- Keep the magnification of your telescope below 300x.
- Try not to pre-empt what you might expect to see. Do not be persuaded by prior knowledge that you have seen something when really you have not.
- There will always be exceptions to many of these rules. This will depend very much on your local weather and light pollution conditions.

Can you think of anything else?

Even after many years of observing, the author still feels a child-like thrill when he sees something unexpected in the sky, or when he stumbles across something he has never seen before drifting into his field of view. Observing something that you never thought possible can give you a real buzz when you have a really great night out under the stars. Yes, it can be very frustrating when the weather prevents you seeing a long expected astronomical event. But when it does all come together nicely and you have a very successful night, the hobby just cannot be beaten.

1.6   Managing Your Expectations

The media certainly does not help in many aspects of its reporting. The Venus transit in 2012 was heralded as a once-in-a-lifetime event. (if you did not see the one 8 years previously). In their words, it was going to be "Spectacular". This raises people's expectations to an extremely high level. Many people may have been expecting a huge spectacle and in reality would have observed a black dot moving slowly across the Sun's surface that day (had they used proper solar filtering).

Would they have felt cheated and let down by the sight itself? For an experienced observer this would have been a very interesting experience (had many not been clouded out) and quite exciting to see. An observer who has had many years of experience would know more or less what this event should have looked like long before it was viewed. Many potential observers without this experience, who listen to the media's hype would not know what to expect and would have had their expectations raised to inappropriate heights. The experience of the observation would have fallen far short of expectations and been disappointing. Experiences like this put off many potentially good observers before they have really started to see things properly.

Most media reporters, unless they are a specialist in science and astronomy, are not experienced enough to digest the full facts thrown at them when an astronomi-

Fig. 1.6 The 2008 transit of Venus (Courtesy of the author)

cal story hits the headlines. So reports are often flawed with inaccuracies. How many times have you found the word astrology substituted for astronomy in space news? Far too many! So take a lot of the information given from within popular reporting circles fairly lightly. They frequently announce these things well after the event so always try and get your up to date information from sources more dedicated to astronomical observing and information sources. If you need to approach the media to announce a story, always be extremely careful and make sure that you choose your wording with extreme care to avoid misinterpretation. So easily can your information appear somewhat distorted in the final article so that it fits a reporter's hunt for a "sensational story".

A this point it is worthwhile, before we start discussing telescopes and instruments, to talk about your eyesight and the implications it can have on your observing.

1.7   How the Human Eye Works

Let us take a look at the sensory organ that allows us to enjoy the hobby. The eye is a wonderful feat of biological engineering that has developed over millennia.

Incoming light is focused by the outer layer of the cornea and the lens coming to a focus onto the back of the eye where an upside down image is formed. The light sensitive part at the back of the eye is called the retina. This has a large number of light-sensitive cells on its surface, which pick up photons of light changing it into electrical energy so it can be transferred via the optic nerve onto the brain. There are two types of cell present in the retina. Rod cells are sensitive to light and Cone cells, which are sensitive to both light and detect color. Each cone cell is connected to the optic nerve via a single nerve fiber. The Rod Cells are connected in groups

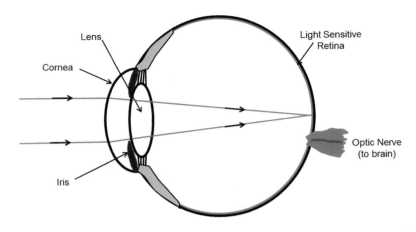

Fig. 1.7 Diagram of the structure of the human eye (Courtesy of the author)

to a single nerve fiber and are the most numerous type of light sensitive-cell present. Cone cells are more densely clustered around the center of our vision. This is where we see the most detail and color.

1.8   Dark Adaptation

The amount of light entering the eye is controlled by the Iris. This is the ring of muscle just under the cornea that gives your eye color. In bright light, circular muscles contract to make the pupil smaller (Much like a draw-string closes a school sports bag). In dim light radial muscles contract to make the pupil bigger. So by constant adjustment the correct amount of light is allowed into the eye depending on the lighting conditions.

When observing in the dark, the pupil is usually as wide open as it can get. This is to let as much light into the eye as possible. Becoming dark-adapted and seeing in the dark is a much longer process than just the pupil changing in size. The change of pupil size is almost instant, but when you turn off the light at night, it takes quite a while before you start to see things in the darkened room. There is something else going on. Your eyes take time to adjust and become properly dark adapted.

Chemicals within the rod and cone cells are much less sensitive to light when they have been exposed to light. When light hits any of the cells on the retina a chemical change occurs. This will eventually produce an electrical current to send to the brain. The chemical change that takes place bleaches the photopigments essential for the light sensitive cells to work. It can take up to 30 min for the pigments in the cells to regenerate after exposure. In rod cells, the photopigment is called visual purple and becomes almost transparent when exposed to

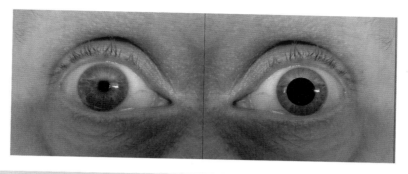

Fig. 1.8 Diameter of pupil changing with different light levels. Brightly lit left, in darkness on the right (Courtesy of the author)

light. Photoreceptor proteins called Photopsin (Iodopsin, or Cone Opsins) is present in cone cells. There are different classes of photopsin present in cone cells, which absorbs a different wavelength of light. This determines the color sensitivity of the cone cell; Red, Green or Blue. L-Cones absorb Red light, M-Cones absorb Green light and S-cones absorb Blue light. L=Long, M=Medium and S=Short (Wavelength). Therefore, the combination of the three types of cones enables us to see in color. One note to add, which is very relevant when observing ionized nebulae is that rod cells are more sensitive to blue light and respond extremely poorly to red light.

There is a well-documented effect called The Purjinje Effect, which affects how we see red objects. Our eyes are in fact very poor at picking up the color red, especially at low light levels. In bright lighting conditions this isn't too much of a problem, but when light levels are low, as is always the case when observing the sky, the cone cells in our eye are much more sensitive at the blue/green end of the spectrum. So red objects seem a lot fainter than they should otherwise appear. This is why we struggle to see many of the hydrogen alpha regions in the sky as they shine very weakly at the red end of the spectrum. This makes it even harder for us to see ionized nebulae. But it does enable us to use a dim red torch in order to preserve our night vision.

The longer you stay in the dark, the more the photopigments regenerate and the more sensitive the cells become to light. As the cells become more sensitive you will be start to see fainter objects. As the cone cells are less sensitive at low light levels what we see at night is mostly due to the rod cells detecting the light. Therefore in dim light conditions we see things in monochrome.

The moral of the story is. If going out and doing some observing, try and keep in the dark and stay that way for at least half an hour. Stay out of the light, don't look at bright TV or computer screens, or camera screens. Only by keeping your eyes in relative darkness will your dark adaptation become complete. Your eyesight will then be in its best condition for observing faint objects.

1.9   Factors That Can Affect Your Night Vision

There are a number of factors that can affect the sensitivity of the eyes preventing you from seeing the fainter objects.

Tiredness

Being tired will severely affect your observing. Unfortunately the nature of the hobby quite frequently means you will need to stay up late, or even observe right through the night to see what you want. The faintness of objects you will be able to see, as well as your concentration and judgment will certainly suffer if you are too tired to do justice to the hobby.

Vitamin Deficiency

Vitamin A is an essential ingredient for use in your vision. Lack of vitamin A will certainly make your night vision less sensitive. So make sure that you have a well-balanced diet.

Illness

If you are suffering from a cold or any other sort of illness, this will indeed affect the sensitivity of your eyes. If you have got an illness of any kind, then you really ought to ask yourself what you are doing observing out in the cold anyway. You really won't do your health or your observing confidence any favors by going out observing when feeling under the weather.

Alcohol

Drinking alcohol before observing will definitely prevent you seeing fainter objects. It has two effects, it stops the pupils opening wide enough to let in the maximum amount of light. The sensitivity of the rods is also impaired and if you drink too much then your vision can become blurred. If you do drink before observing, take it in moderation. Another thought to add is that if you do drink alcohol and operate an expensive telescope, it will increase the chance of having an accident.

Caffeine

If having a long night out, staying awake can sometimes be an issue. Drinking coffee or tea seems to be the obvious choice of drink to combat tiredness. Intake of caffeine does reduce the sensitivity of the eye at low light levels. So as in all things moderation is the key.

And don't forget, the more liquid you consume, the more trips to the bathroom will be required.

Smoking

Regardless of the detrimental effect on your health from smoking, literature seems to be confused as to whether smoking impairs or enhances your night vision. Reduced oxygen in the smoker's body and increased levels of carbon dioxide in the blood should reduce your eyes sensitivity. Many studies have shown that there is some enhanced light sensitivity. This may be due to the stimulant effect of the nicotine contained in cigarettes.

Drugs

Observing while under the influence of even prescribed medication will affect your vision in one way or another. This will of course depend on the type of drug being used. Again, if you are on medication are you really fit enough to be staying up late and being out in the cold observing? Many prescribed drugs can also make you drowsy. Not an ideal combination especially when staying up late.

Holding Your Breath

While observing, especially on a cold night, the moisture in your breath can quite often fog up your eyepiece. Holding your breath will prevent this. When you do hold your breath, the levels of oxygen in your bloodstream are reduced slightly. This will produce hypoxic effects and the sensitivity of the eye to light will be reduced. So holding your breath while observing is a hindrance to seeing faint objects and isn't advised.

1.10   Maintaining Your Dark Adaptation

Now you've spent almost an hour getting your eyes as sensitive as possible, don't spoil it! The rods in your retina are far less sensitive to red light and your dark adaptation is affected less by it. If you do need to look at something in the dark use a dim red torch, or flashlight. Make it just sufficiently bright enough so that you can just read what you want to with it. If you haven't got a red torch you can adapt an existing one. This can be done in two ways:

Obtain some red cellophane from your local craft or stationary shop. Cover the front of the torch with a few layers of cellophane to reduce the amount of light and give it a nice dim red glow.

Alternatively you can use red nail polish to coat the front of the torch. Two or more coats may be required to reduce the amount of light to an acceptable level.

A head torch with a red LED light is very useful for hands-free operation. Some of them are very bright unless you can dim them by painting on a coat of nail polish or cover with layers of cellophane.

The other thing to remember about a head torch is to ensure that they do not shine in other observers eyes when you are talking to fellow observers in the dark while observing. The eye may be less sensitive to red light, but shining a bright light, even a red one, into someone's dark sensitive eyes isn't going to make you very popular.

An eye patch can be used to cover one eye while using a torch. This will retain the night vision in that particular eye. These are also useful to cover the spare eye while observing, so you don't have to squint or close your spare eye to keep out extraneous light.

Laptop/Computer/Phone

If you are using electronic equipment while observing, avoid having a bright screen if at all possible. Planetarium software, or other software packages have a night vision mode, which turns the graphics red. There are also software packages that can be installed on your laptop that will reduce the brightness of the monitor. Even when some of these methods are used, the screen can still be bright enough to ruin your dark adaptation. Many models of laptop enable you to dim the brightness of the screen. This is often within the computers power settings. So make the screen as dim as possible and use any night vision mode available as well. You could also attach a layer of red cellophane over the screen.

Lighting Cigarettes

The brief flare of bright light from a lighter or match whilst lighting a cigarette is enough to destroy your dark adaptation.

Observe Fainter Objects First

It's no use trying to look for a faint galaxy if you have just been looking at a bright planet. The brightness of a planet will destroy your dark adaptation. So if possible observe bright objects once you have finished hunting down your fainter objects. But don't forget to observe some bright objects while the sky is still darkening.

1.11   Eyesight Correction

Eye Glasses

If you wear eye-glasses, your prescription will affect how you are able to observe. Removing the eye-glasses will dramatically improve the image you see in the telescope, cutting down on reflections and making it much easier to hold your eye in the correct position. Nevertheless, not everyone will be able to do this.

People who only suffer long and short-sightedness can take off their eye-glasses when they are observing. The difference in the eyes focus can be accommodated simply by adjusting the focus of the telescope. The view they have through the telescope will be as good as someone who doesn't require eyesight correction.

Unfortunately, if a person suffers from astigmatism, this cannot be corrected by focusing the telescope. In this case, their eye-glasses or contact lenses will still need to be worn when observing through the eyepiece to get the clearest view. If you need to wear your eye-glasses, you will need to make sure that your eyepieces have a long enough eye-relief (See eyepieces later). Some eyepiece adapters have been manufactured to correct astigmatism while using a telescope, but these are an expensive solution.

Laser Surgery

Laser surgery can correct astigmatism and short and long-sightedness. In this case no eye-glasses will need to be worn following successful laser surgery as the eyes have been entirely corrected. There is a lot of controversy surrounding laser surgery and whether the cornea does transmit as much light after treatment. As with most surgery there are risks, but with the correct treatment many observers have seen great benefit from the procedure. The corrective surgery can in some cases be very expensive. This will depend on the amount of correction that needs to be applied to correct the patients eyesight.

For those considering laser surgery to correct their eyesight, there is not a great deal of literature published regarding laser surgery and its effect on astronomical observing. A good review, but now sadly outdated was published by Christopher Wanjek in the September 2009 Sky & Telescope, issue 36, under the title "LASIK: Eye Surgery for the Amateur Astronomer". For the author's own account of his laser surgery experience, you may view the story at www.eagleseye.me.uk/Laser.html.

1.12   Set Yourself Realistic Targets

Observing takes patience and practice. It takes a lot of observing experience to develop your "seeing eye". You can only make this progress by properly planning what you want to do and setting yourself realistic targets. However, what is realistic? Do some scouting around. Find out what other observers are achieving and read a variety of observing books to get even more information. Make friends with your local astronomy club and discuss with others what they are doing. Find some objects they have seen that you will be interested in seeing yourself. Be determined to "give it a go". You will find there will be some people who will try and discourage you, persuading you some things are just far too difficult. However, persevere, negotiate your way over the many setbacks and many a time you will prove them wrong time and time again. Now that we are in this frame of mind, let us get going with our next written exercise:

Exercise 1C: Make a List of Observing Targets

Make a list of 30 targets throughout the coming year that you would like to observe.

Refer to annual handbooks and calendars to identify upcoming events you might be interested in that are visible in the coming months and add those you are interested in seeing to your list as well.

Divide these objects into observing seasons. Spring, summer etc. Full Moon, Half Moon etc. So that you have a good idea when these objects are best viewed. Add them to a diary or print out your list to consult when needed.

As stated before, be realistic when setting your targets, but at the same time, also include one or two objects that are a little bit more challenging. Some of the objects you put on your list now may not be achievable right this minute. This list can be used as a handy observing plan to guide you around the maze of objects on offer throughout the year. Do not be discouraged if you fail to find your targets the first few times. By being a little more focused while observing you will gradually pick your way through your targets. As some of your easier objects are ticked off your list, your confidence and "seeing eye" will really start to grow. Eventually you will find you will be able to start ticking off some of those more challenging objects and will soon start to add other more challenging sights to your ever increasing list of achievements.

That is the beauty of our hobby, you can continually develop and grow and you can get as involved (or not) as you like. Added excitement is also never too far away as you just never know when something new might suddenly appear in the night sky to take you by surprise, be it a bright comet or a supernova, for example.

1.13   Evaluate Your Favored List of Targets

Now you have now got a list of objects, have a good hard look at the list you have just written. What type of objects have you included? Are there lots of planets? Do they include lots of lunar features? Are they mostly deep-sky objects?

Your list will reflect the types of objects that you are really interested in observing. Maybe you are particular in your interests, only concentrating in a particular field, or like the author just love seeing anything up there that is visible at the time of observing. Perhaps you can already see that you are already on your way to perhaps specializing in some way, and starting to make plans for your future observing sessions. This list will help concentrate your thoughts when observing and focus your efforts when out under a dark sky. This observing plan will increase the chances of achieving your goals and managing your expectations. As you work through your list, add more objects. It really will help you concentrate your efforts. If you follow your passion you will soon find yourself tackling objects you initially thought were well beyond your current telescope setup or capabilities.

1.14   Assess Your Abilities

When you do start to carry out more detailed observing, and take a bit more time over it, step back and have a look at the results. Have you observed this object before? How did the view differ from when you observed it previously? Was your new observation better or worse than the previous one? If it was different, can you pinpoint what it was that made that difference? Was it your observing skills starting to develop? Did you use a different technique, telescope, eyepiece or conditions? What have you learnt from your experience? Can you identify why your observation was a success or failure? What could you do to improve your observing or improve your observing experience?

A big temptation when you ask yourself this question is to try and avoid answering with "Buying more accessories or upgrading my telescope", at least initially. Get out observing and get the most from the instrument (s) you already have. Really get to grips with its capabilities before you start investing more money in more or different equipment. Only by knowing your equipment properly and getting to know it inside and out, will you really start to get the most from it. In Chap. 3 we will turn our attention to the equipment you already have and try and identify if there is anything about your current setup that puts you off getting out and observing.

Chapter 2

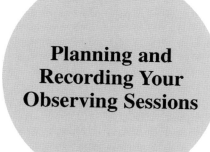

Planning and Recording Your Observing Sessions

2.1 Preparing for an Observing Session

After what seems like weeks of cloudy skies, you suddenly find yourself faced with a clear night. All too often, the keen observer rushes out totally unprepared and has little idea what they are going to observe. That is all well and good for short impromptu sessions, but if you are going to make a good night of it, you will be much more productive if you do prepare properly and do your homework.

2.2 Weather

What is the weather going to do tonight? Check the weather forecast. Is that current clear spell going to stay for the whole night, or is it just a short-lived event and clouds are soon to arrive? There are a multitude of Web sites or mobile phone apps available for checking the weather or looking at the latest satellite image, so make sure that you consult them before you start setting everything up. It can be frustrating and disheartening to take the time and effort to set everything up. Then, just as you start enjoying the sky, the clouds roll in and spoil your night. There are a number of generalizations that can be made about the weather and which should be kept in mind, wherever you are in the world.

Cold Air

Skies are usually very transparent, but the seeing is generally very unsteady. Good for deep sky objects, it is not so good for observing planets and the Moon.

D. Eagle, *From Casual Stargazer to Amateur Astronomer: How to Advance to the Next Level*, The Patrick Moore Practical Astronomy Series, DOI 10.1007/978-1-4614-8766-1_2, © Springer Science+Business Media New York 2014

Warm Air

Some cloud, haze or dust may be in the atmosphere. Here the transparency is usually quite bad, but the air is quite still and seeing very steady. This can be great for observing the Moon and planets, but awful for seeing faint deep-sky objects.

Solar Observing

On clear days observing the Sun make sure that you wear suitable clothing and cream to protect you from the Sun's UV. Even on a cold day, the radiation from the Sun can easily burn your skin.

Insect Bites

Observing at night can expose people to the worse of the biting insects. There are many products available to help protect you, but they are often oily and extremely smelly. Many of the repellants contain a substance called N,N-Diethyl-meta-toluamide. (DEET). This should not be used on damaged skin and has some potential health concerns. It is also an extremely effective solvent and can dissolve plastics, synthetic materials and varnished surfaces, especially when used at more concentrated preparations. So it really isn't a very good idea to get DEET anywhere near your precious observing instruments and accessories.

Winter Observing

It can be extremely cold on clear nights. Wrap up with many layers of clothing. What you will soon find out when you start observing is that the temperature drops rapidly once the Sun goes down. First and foremost, think comfort. The best seeing conditions are very often on crisp, cold and frosty winter nights. It is very easy to get lulled into a false sense of security when you first go out observing. The temperature might feel quite comfortable initially, but as the evening progresses, or the early hours approach, under a clear sky the heat rapidly dissipates. It is very easy to get cold. Once you are cold, it is extremely difficult to get warmed up again. If you are not comfortable and warm, your concentration will suffer.

If you are lucky enough to have an observatory, it might shelter you from the wind somewhat, but it still gets very cold. Any heating inside would produce turbulent currents out of the roof and this will destroy the seeing. Some lucky observers have a warm room they can retire to when it gets cold. The author's observatory is so close to the house that he frequently goes back inside to warm up.

Many a great observing session has had to be abandoned when the observer starts shivering, feels totally miserable and just cannot stand the cold any more. Once you get to this stage, it's game over. The only way around this is to wrap yourself up as much as possible to retain as much of your generated body heat as possible.

The secret to keeping warm is to wear many layers of clothing. If you get too hot, you can always take off one of the extra layers. Wear warm undergarments. These are extremely useful at keeping the body warm on very cold nights. They act as a very good base for keeping in your body heat. Wear thick socks and stout boots. The feet are quite often the first part of the body to start suffering when the toes go numb. This is especially so if standing on damp grass and your feet get wet. Also wear a thick coat. One with many (clean) external pockets will be useful as you could put your hands in them to keep them warm, or keep your eyepieces in them. This will also stop them misting up.

Wear gloves, but the thicker the gloves, the harder it will be to operate equipment. It might be useful to have a pair of gloves where the fingers can be folded back when required to change eyepieces, operate a camera etc. Hand warmers can also be very useful to warm up the hands. They are available fairly cheaply from camping and outdoor shops. They are small pouches that contain a gel. There is a metal clicker inside which the observer operates to release the heat. The gel rapidly crystallizes, releasing heat in the process. These can be placed in pockets for warming the hands. They are re-generated by boiling in a saucepan.

Also wear a hat, and the woollier, the better. Do not forget, in the dark no one can see what you look like, so do not worry if the style is not "in fashion". As long as it keeps you warm, it means you will be able to get the most from your clear skies.

There are still going to be some occasions where the worse of the cold will always win, but the best-prepared observer will be able to get stay out longer and be able to achieve far more.

2.3   Gain Knowledge of Your Quarry

When hunting, knowing the behavior, intimate details of their nature and the habits of your quarry will always give you the best chance of a successful hunt. The more you know about them, how they behave, how they might look, and when and where are the best places to find them, the greater your chances of success. Astronomical observing is very much the same. The observing targets are your quarry, now we need to find out how best to track them down and observe them.

What Objects Are Visible Tonight?
As we will see later in this book, every object has its best time of the year or night for viewing. The best time to view most objects is to catch them as high above the horizon as is possible. Having good knowledge of the objects, and when they are best observed, will increase your chances of observing them successfully. If you timed your observing session properly and went out a little earlier, or a bit later, would it make your intended target(s) easier to observe? Do your homework before stepping outside.

Check for predicted or new events. Is there anything out of the ordinary or unusual happening tonight while you are out observing? We have all done it! Many a time an observer has had a great observing session out in the garden and then read the next morning of other observer's experiences of an event that they too should have seen had they been looking in the right direction that very same night. You just never know when a supernova is going to go off, or a new comet has been discovered. Only by keeping up to date will you know what you need to look out for. Keep your eye out in astronomy magazines, Web pages, astronomy forums and even the national news, so you are as prepared as you can. Just try not to notch up too many missed opportunities.

2.4 Keep an Observing Notebook

When making your observations record everything by writing down what you have seen. This produces a permanent reference for future use.

Some people diligently fill in dairies every day, documenting down to the finest detail everything they (and in some cases what other people) do. Others find the whole process a real drag and never fill one in. In this case, they rely solely on their memory to recall what has happened.

Observing is a constant challenge. What you have seen should, in an ideal world, always be recorded in some way. From the constant bombardment of information we are exposed to today, your memory can, over time, often be tricked into convincing yourself that you have already seen something that you really have not. Many experienced observers are the worse for not documenting completely what they observe, the author being a typical example. Looking in your completed observing log in the future will tell you exactly what you saw and how you saw it. This is especially useful if going back and revisiting previously observed objects.

If you do try and find one of your targets and fail, make sure that you also include that in your notebook. Failures in finding an object, or failing to see some particular detail will help to develop your observing skills. So always include your failures in your write-ups alongside your successes.

How detailed should your recording be? That decision is entirely yours to make. There really is no right or wrong way of doing things and there is no need to be too regimental about the whole thing if you do not want to. Remember that it is your hobby. You do it in the manner that gives you the most pleasure. Put too much pressure on yourself, or try and do something you are not really comfortable in doing and you will not enjoy the experience. This will result in the hobby becoming a drag and you will discourage you from going out and observing.

As a minimum, keep a small notebook handy to use as an observing log. A typical observing entry in your log should, as a minimum, include the information in Sect. 2.5.

2.5 Date and Time of Observation

Date

If an observation is to be submitted, it will usually be dated according to Universal Time. So make sure that your conversion to Universal Time (and Date) is correct to give you the correct time and date. Especially when observing at midnight.

Gregorian Calendar

This civil calendar is the one we recognize and share across the world today. It is composed of the Day, Month and Year. It is commonly expressed in the following format: dd/mm/yyyy, or mm/dd/yyyy, where numbers replace the letters accordingly.

However, there is a need to watch out here, especially if sourcing information from the internet. The convention of writing down the date differs in other countries. So the 15th of August 2013 will be expressed in different ways. In the UK and Australia it will expressed as 15/08/2013. In the USA this will be expressed as 08/15/2013.

Luckily on days in the month above 12 the difference is easily spotted. Let us take another example of 8th March 2013. In the US it will be written as 08/03/2013. In Europe and Australia it will be written 03/08/2013. So the dates can be easily confused if you interpret the wrong format.

This is something that will be easily mistaken if you have electronic equipment, such as telescope handsets. They often only accept one form of the date, which may differ from your local tradition. Some reports require you use yyyy/mm/dd when submitting observations. So make sure that you are aware of this. Keep your eye out for any differences that could cause confusion.

The Julian Date

Although the more casual observer rarely uses it himself, the Julian date is used quite a lot in astronomical calculations, so it is worth the observer being aware of it. These times can be obtained from most planetarium programs, other software or online. This date is derived from the time that has elapsed since midday on January the 1st 4713 BC.

If you are interested in why this particular date was chosen, visit this Web site: http://scienceworld.wolfram.com/astronomy/JulianDate.html

The Julian Date is quite often used by planetarium software in its calculations to avoid the annoying complexities of extra days added onto leap years and the like. Using the Julian Date, midday on our chosen date of 15th of August 2013 will be expressed as the Julian Date of 2456520.00000. The five decimal places enables the observer to break down the day into very small fractions, so that a very precise (perhaps too precise) time of a particular event can be given. Using the Julian date also avoids the confusion if a change in date occurs while making observations.

Time

Universal Time (UT) is generally used in astronomical observations for consistency. This is equivalent to Greenwich Mean Time (GMT). For convenience to us humans the Earth has been divided up in a number of time zones. This is so that at midday the Sun is always approximately at its highest in the sky, wherever you are on the Earth. At the extreme north and south the Sun may not be above the horizon in the depths of winter. However, it will be at its minimum distance below the horizon at that time.

A diagram of the Worlds time zones is available at http://www.worldtimezone.com.

This diagram is a simplification as there are more than 24 time zones that can be seen at first glance. There are some isolated pockets of local time within other time zones.

As an observer, you probably know in which time zone you reside and how many hours ahead (East) or behind (West) Greenwich (UT) you are. Therefore, an observer living on the Eastern Seaboard uses Eastern Seaboard Time (EST) and this time zone is 5 h behind UT.

Daylight saving time comes into force in many countries from spring until the fall. This usually puts the clocks an hour ahead at your location for the summer months. Do not forget to take this into account if daylight saving time is in operation in your country at the time of your observation. Convert back to Universal Time, unless you state explicitly that local time has been used.

A convenient Time Zone Converter can be found here:
http://www.timeanddate.com/worldclock/converter.html

Sidereal Time

This is a measure of the time based on the Earth's rotation in relation to the stars. This is measured by noting the line of Right Ascension that is moving across the observer's meridian at the time of observation. i.e. the time that has elapsed since the vernal equinox has passed across the meridian. This is equivalent to the right ascension of a star that is currently on the meridian. This may or may not be included in your observing log, but it is useful to be aware of your current sidereal time.

Sky Conditions

Make a note of the general sky conditions while making your observations. Is the sky really dark, observations made in Twilight? Is it slightly hazy (haze generally helps with planetary or lunar observations)? Are the skies really clear and transparent? Is there any wind? Are the stars twinkling wildly or do they show a much steadier light than usual? How much light pollution is there at your observing site? Is it better or worse than normal? Are there any neighboring lights or general light pollution?

It is always worthwhile checking out the general sky conditions as you might just spot some auroral activity or the presence of noctilucent clouds while doing so. If you are really lucky to have very dark skies, the Zodiacal Light or the even more elusive Gegenschein may also be visible. Therefore, it is always best to keep a look out at all times and always have a broader awareness of the sky as well as your intended target.

A night where the image is steady and calm would be perfect for viewing double stars, planets and the Moon. On nights of very poor seeing, nothing but the most casual observations can usually be made despite the fact it is a perfectly clear night, with the image shaking and wobbling all over the place. On nights like this fine detail is extremely hard to resolve.

2.6   The Seeing Scale

Astronomers across the world use the Antoniadi Scale to give an indication of seeing conditions while observing. This is caused by the turbulence of the air in the atmosphere above our heads. Conditions can change in a matter of seconds. A perfect night for viewing will have naked eye stars that are visibly twinkling a lot less than usual. This is usually very obvious if you observe a star, the Moon or a planet through the telescope.

The seeing scale is a five-point system as shown below:

I. Perfect seeing, without a quiver.
II. Slight quivering of the image with moments of calm lasting several seconds.
III. Moderate seeing with larger air tremors that blur the image.
IV. Poor seeing, constant troublesome undulations of the image.
V. Very bad seeing, hardly stable enough to allow a rough sketch to be made.

This scale can sometimes be very subjective. Many an observer has frequently given seeing conditions a score of II or a III, but when they have gone to clearer and better skies, found others referring to II and III, with much steadier skies. Therefore, their system was much different to others. This was because of their own personal experience. Whatever your perceptions, always try to be consistent in the way you measure these things, so you can compare notes later. If you failed to see something on one night because the seeing was so bad, but you saw it another night when the seeing was better, your notes should indicate why that observation was successful. You will then start to appreciate the best conditions that will start to give you the greatest opportunities to observe your intended objects.

Another scale sometimes used to measure seeing is the Pickering Scale. This uses a 1–10 scale with one being very poor seeing. Ten having excellent steadiness to the image with perfect observing conditions.

2.7 Note Down How and What You Observed

What Object/s Did You Observe?
A planet, Double Star cluster, Nebula, Galaxy or another object? Make sure you use is currently accepted name, e.g. Messier, NGC catalogue etc.

What Optical Instrument Did You Use?
You might have made an observation with the Naked Eye, Binoculars or a telescope, or a mix. If a telescope was used always include its Aperture, Focal Length and whether or not it was driven to track the sky. Did the object(s) look different in different instruments/apertures? What eyepiece(s) and magnification did you use? Did you use a variety of eyepieces to try and get a different view? Did the view differ? How? Were any filters used to aid your observation?

Did you swap filters to see if it made a difference?

Did you try anything else differently?

Your Observations: What Did You Actually See?
Did you notice anything different or unusual about the objects you observed? Did you manage to see any detail in the object? Did the light intensity vary across the object or was it smooth? Did the view differ when observed directly

or when you used averted vision? From what you tried, what worked really well? What didn't? How different was the view from the different combinations you tried?

2.8 Astronomical Drawing

They often say that "A picture is worth a thousand words" and so it is. Making a quick sketch of what you have seen really does capture what you have seen while observing and is an excellent way of reminding you what you saw while you were "out there".

Before the advent of photography the only way to produce an accurate visual image through a telescope was to draw what you could see. Eminent names in astronomy including the likes of Christian Huygens, Giovanni Domenico Cassini and William Herschel all made significant contributions to our knowledge of the solar system through their drawing work. The Dutch astronomer and mathematician Christian Huygens produced the first known drawing of the Great Orion Nebula in 1659. In doing so he noted that the interior of the nebula consisted of stars – a first for seventeenth century science.

Today, astronomical drawing is a purely recreational activity. While no longer directly contributing to our developing scientific knowledge, it does challenge our senses – particularly our vision, spatial awareness and hand-eye coordination. It's very accessible too. Anyone can draw. You don't have to be an artist. In fact the less artistic ability you have the better. The whole process needs to be objective and methodical, relying more on your ability to push the limits of your own vision than anything else. As always the case in visual astronomy, the more you look and observe, the more you will see.

Why make drawings? Visual astronomy is about the experience. It's the thrill of tracking down a distant object many light years away and seeing its photons for the very first time. Perhaps you're seeking out minute or low contrast details that you've often read about but never seen first-hand. Sometimes it's the faintest of smudges in the night sky that mean so much when you know their enormous distances and the stories behind them.

Only when you start to understand what you're looking at, and begin to appreciate the enormity of scales involved across dimensions of space and time, does visual astronomy come alive for a lot of people.

In addition, it's right here where astro-drawing develops meaning. We're not talking about the aesthetics of a pretty picture, but more about the raw power of emotion in capturing what you see. If follows that you might also like to relate and share your visual experience with others. We all want to share with others what excites us the most and astro-drawing gives each and every one of us that outlet. All that is needed are a few pencils, some paper, a dim red light and plenty of patience at your telescope.

2.9   Getting Started with Drawing

The starting point for every drawing must be the quality of the visual observation. The very best astro-drawers are people who train their eyes to look for small-scale tonal variations in the objects they track down. Like anything it does take practice. The more you observe and practice seeking out these small variations, the more detail you'll soon find yourself able to see and record in your drawings.

Perhaps the most important aspect in all of this is having the knowledge to stack your visual observing conditions in your favor. Without doubt, the more your eyes are in tune with the apparent brightness of the object you're observing, the more detail you will get out of it.

2.10   Equipment Needed for Drawing

Drawing is a very low cost imaging route. There's very little in the way of financial outlay to get started. All that you need is:

1. Pencils – A set of graphite pencils from your local stationery store will do the trick. To give you the freedom to plot those small-scale tonal variations that you might see in some objects it's best to invest in a set that offers different graphite grades. A set that ranges from 4H to 6B will work well.
2. Colored Pencils – A set of colored pencils are useful to have but not essential. More than 95 % of the objects that you view in your telescope will be seen by your eyes in grayscale tones. The other 5 %, which includes, Mars, Jupiter, Saturn, some double star systems and a handful of nebula, will show varying degrees of color to the human eye. For the planets colors will mainly be earthy tones – browns, oranges, yellows and maybe some green. For stars and nebulae shades of blues, yellows, oranges and reds might be needed.
3. Sketch Pad or Paper – When starting out a 90gsm or 100gsm standard copier paper attached to a clipboard will work well. As you grow in confidence its worth experimenting with different textures and weights of paper to see if it adds to your drawing style or detracts from it. Don't forget that more flimsy paper will be the first to suffer in damp weather.
4. Drawing template – To get an accurate drawing you will need a frame of reference. If you're drawing a whole eyepiece view then a printed or drawn circle representing the outer edge of your field of view is recommended. The best way to do this is to set up a template containing varying sized circles on a word processor or similar software. You can then experiment with different sizes for your eyepiece frame of reference to see what works best for you.

Drawing planetary detail requires a template more representative of the planet's visible disc. Many of the observing sections of astronomical organizations discussed later have drawing templates and observing forms available for planets and deep-sky observing. It's worth noting that all the planets but Saturn have a standard template that can be used for any observation of the planet in question at any time when it is visible in the night sky. The angle of Saturn's rings as seen from our viewpoint here on Earth changes from year to year and even from month to month. You will therefore need to select a template that is best representative of the angle of Saturn's rings at the time of viewing.

A number of planetary (and other) astronomy drawing templates can be downloaded from here: http://www.perezmedia.net/beltofvenus/templates.html.

Red Light

Drawing what you see on paper of course requires some light for you to accurately plot down the details. Don't forget to use a dim red torch to enable you to see what you are doing, but still preserving your night vision.

2.11 How to Draw

Once you're confident that your eyes are fully dark adapted and you've spent enough time studying the object in search of those tiny standout details that characterize it, you are now ready to put pencil to paper.

Drawing Galaxies, Nebulae & Globular Clusters
The best way to draw galaxies, nebula and globular clusters is to use grey on white, where the brightest parts of the object are represented by the darkest shades of grey.

1. Start by plotting down the brighter stars that you can see in your field of view. This will help you pin down a framework in which to draw your object. It's quite surprising how easy it is for the eye and brain to exaggerate size and distances when drawing from the eyepiece. By building a framework of reference stars

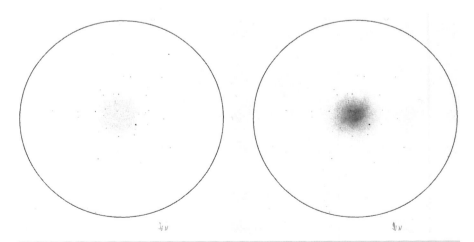

Fig. 2.1 Draw in the bright stars first, before putting in the fainter ones (Courtesy of Seb Jay)

inside your field of view you'll be able to contain your drawing better so that it is representative of actual viewing size.

2. Using a hard pencil (4H or 3H) held at an oblique angle to the tip lightly lay down a patch of graphite on your paper that represents the shape and maximum extent of the object that you can see. Use a finger or a piece of cloth and lightly spread the graphite around in the area. This will help you get a smooth and fuzzy look to your object, especially around the edges as it fades to the background.

3. Using progressively softer and darker pencils, build up layers of brightness visible in the object. Galaxies and globular clusters normally brighten to their core, so often it is a matter of building darker shades of grey towards the central area of the object, obviously being mindful to develop any lighter and darker features visible as you go. Drawing nebulae can be a bit trickier as frequently there is no central condensation. Therefore more care is required in building up the brightness layers.

4. Complete your drawing by filling in the fainter stars that you can see in and around your field of view, plus any fine detail that leaps out at you during this process. Often, when you're concentrating on pulling out the faintest points of light more detail in the object you've drawn suddenly jumps out at you!

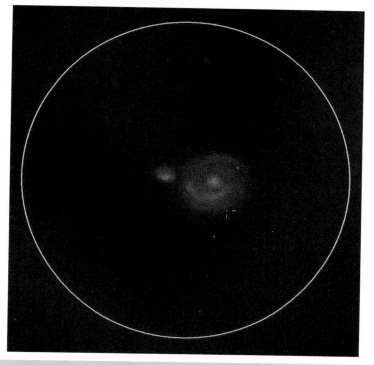

Fig. 2.2 Drawing of the galaxy M51 and Supernova 2011DH (Courtesy of Seb Jay)

Drawing Star Clusters

As for galaxies and nebulae, star clusters are best drawn black on white.

1. Start the same way as above by plotting the brightest stars in your field of view to generate a reference frame for the rest of the drawing. The very nature of a star cluster means there will be a larger number of brighter stars to note down at this initial stage than in a drawing of a galaxy or nebula. The key to plotting these bright stars accurately is to look for patterns – particularly triangles.

   Triangles in the broad sense of the term are very useful for judging relational positions of stars. Once you have two stars accurately plotted in a line your third star will always form an apex to a triangle (normally a scalene triangle where no side or angle is equal) when using the other two stars to form a baseline. Every subsequent star then will also form an apex to a triangle when used with a choice of two other stars on your developing drawing.

   As you note down more stars you might find it easier to switch to looking for other patterns that consist of four or more stars. Squares, rectangles, rhombus and rhomboids are among some of the common shapes to look for to help you plot down correct star positions.

2. The brighter your star the softer the pencil you should use. In a range from B to 4B, the 4B is your softest pencil. A 4B pencil lays down graphite the darkest and should be used for marking out exceptionally bright stars. Pencils graded 3B, 2B and B should then be used to plot down stars with progressively less luminance.

   Note that brighter stars tend to shine with a larger halo of light than fainter stars. It's a visual effect and probably needs to be noted on your representation. Therefore, draw brighter stars a little larger than fainter stars to capture this dynamic representation.

3. Fill in the remainder of the cluster's stars moving from brightest to faintest. For very faint stars consider using a light touch with the tip of a 2H or 3H pencil to ensure good brightness correlation with the other stars.

Drawing the Moon and Planets

Drawing planetary detail requires an altogether different approach from Deep Sky objects. First off you're drawing in positive light – i.e. dark features are drawn in darker shades of grey or color and light features in lighter shades of grey or color.

1. Start your drawing by getting familiar with the surface and/or cloud detail you can see on the disc.

   Mars rotates around its axis at a slightly longer period than our 24 h day here on Earth. The features you see on the surface will therefore move through the night due to this axial rotation. Overall you probably have about a 1 h window to plot down features accurately without them shifting too far from their initial positions. Jupiter's rotation is quick. The giant planet completes a full rotation once approximately every 9 h and 55 min. It makes plotting features accurately a little more challenging as appreciable movement becomes obvious after just 20–25 min of viewing.

Fig. 2.3 Drawing of globular star cluster M53 (Courtesy of Seb Jay)

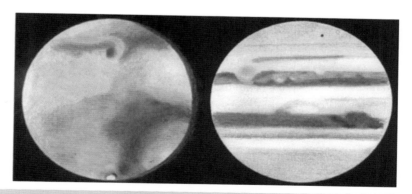

Fig. 2.4 Drawings of Mars and Jupiter (Courtesy of Seb Jay)

2. Note down the positions of the main features you can see. For example, on Jupiter roughly pencil in the positions of the main bands, spots, barges and bays that are obvious to your eyes. Look to finalize the placement of these within the first 5–10 min of starting. These will form the framework for your drawing to which you can relate the finer detail later on as features move across the planets disk.

3. Working from the western limb to the eastern limb, begin building those finer details using progressively darker pencils. Why work from west to east? Just as Earth rotates in a clockwise motion so Mars and Jupiter do the same. Features rise on the planet's eastern limb and set on the western limb. If you work from west to east you can be sure that you'll be able to plot features accurately before they move off the edge of the planets disc.

4. To complete the drawing, spend time at the eyepiece searching out the really fine and low contrast details. During the course of your observation you'll notice that the steadiness of your view will vary, regardless of the overall calmness or turbulence of that slice of upper atmosphere your telescope is peering through.

Most of the time your telescopic image will shake and shudder about and look very smeared; other moments will plateau in a state of calmness, at which point super-fine details on the disc will suddenly be revealed. These moments might be fleeting but you need to be able to make the very most of them to pick out those really small and low contrast features to make your drawing really special.

Once you are happy the resulting drawing is finished it can be cut out and stuck into the appropriate page of your gradually increasingly filled observing log.

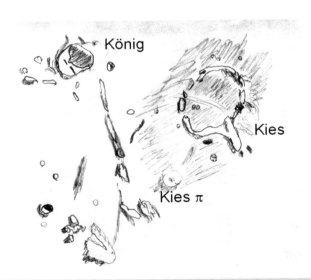

Fig. 2.5 Drawing of the Lunar Surface close to Kies and the Lunar dome Kies Pi (Courtesy of the author)

2.12   Finalizing Your Drawings for Presentation

The best way to do this is to use a flatbed scanner to upload the drawing onto the hard disk of your computer. Unless you have a color element to your image choose to scan your drawing in grayscale as a .jpg or .png file. Use the highest pixel setting as possible without producing too big an image.

Once the image has been digitized there are a couple of other steps you might like to do if you have the appropriate software.

Deep Sky objects may benefit if you turn the image into negative (for a positive light image). That way your dark greys will become light and your light greys dark. The white background will become black. Your drawing will represent exactly what you have viewed in the eyepiece – bright stars and nebulae on a black background.

Once your drawing is complete and you are gaining confidence that your drawings really are representing your eyepiece view, you might want to share it with others. Observing coordinators will be extremely pleased to receive copies of them. Alternatively, how about submitting them to an astronomical magazine for publication?

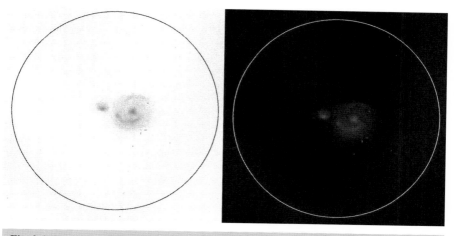

Fig. 2.6 Changing your drawing to white on black (Courtesy of Seb Jay)

Chapter 3

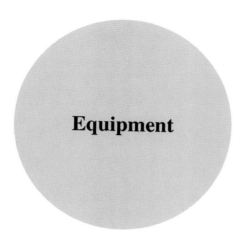

Equipment

This chapter is all about thinking about the equipment you have, or help you identify what type might be useful for the observations you are interested in making.

3.1 Telescopes and Accessories

Having good or bad equipment can make a big difference to your observing experience. So let us start with a look at your existing set up.

Do you find your telescope and other accessories easy to set up and use? Does your equipment show you what you expect to see? What most annoys you about your setup? What accessories do you have? How many eyepieces do you have and what focal length are they?

If you do have frustrations, what is it that makes things difficult when trying to observe? Is it the equipment, your surroundings, or possibly your approach to observing? Can you identify what is making your observing difficult to enjoy properly? A really good rule of thumb is that the best telescope for you is the one that gets taken out and used on a regular basis. Ask yourself the following questions:

Do you take your telescope out and observe regularly?
What equipment do you have?
Why did you purchase this particular setup?
What did you expect to see or achieve while using it?
Have you achieved everything you wanted with it?
Does your equipment constantly fall short of your expectations or exceed it?
Does your equipment frequently let you down in some way?

D. Eagle, *From Casual Stargazer to Amateur Astronomer: How to Advance to the Next Level*, The Patrick Moore Practical Astronomy Series, DOI 10.1007/978-1-4614-8766-1_3, © Springer Science+Business Media New York 2014

Fig. 3.1 A selection of telescopes (but which type really is best for you?) (Courtesy of the author)

Are there any accessories or equipment you bought that you never use?
If so, why don't you use them?
Do you have problems using some of your equipment?
What seems to cause you the most problems when observing?
What could you change to make it work better for you?

At fear of repeating myself, try not to answer this last question with "Buy bigger and better equipment". At least, not yet.

Start to look at your equipment with fresh eyes. Don't forget, just because a piece of equipment is performing well within the manufacturers specifications and at its limits, that doesn't always mean that it will meet up to your expectations. Whatever those expectations might be.

Even a 30″ telescope under pristine skies is never going to give you images as beautiful and colorful as those sent back by the Hubble Space Telescope. Is this what you are expecting? Of course trying to get a relatively small ground based telescope to match a multi-million pound facility orbiting above the interfering Earth's atmosphere is well beyond reasonable expectations. But that is the expectation that some people unreasonably have. You should by now be a fairly seasoned observer and hopefully your expectations would be generally a little bit lower than this. But not too low!

Now you have been using your equipment for a while you should know it really well by now. Do you really? It is a learning curve all the way, sometimes very steep, whatever aperture of telescope or cost of equipment you have. But generally, the more expensive the equipment, the more fiddly it will be to use and the more expensive the repair if you do break something whilst fumbling around in the dark.

Have you had enough clear nights out with your equipment to feel really comfortable and safe using it in the dark? Have you really explored how well your setup really performs? Have you stretched its capabilities as far as possible? It really cannot be stressed enough, have you really managed to get the very best out of your equipment and seen as much as you can with it? Most of us haven't. It can take many, many years to explore all the possibilities that your equipment can reveal to you.

Answer these few questions to see just how well you know your own equipment.

What is the faintest magnitude object you can see with your set-up?
What is the closest double star you can resolve?
What is the most remote object you have viewed so far?
What is the highest "usable" magnification on your set-up, and on what object?

How did you get on?

The only way to ensure that you can answer these questions is to get out and observe as much as you can with your scope. Try some challenging objects to test yours and your equipment's capabilities. OK, so you may not be interested in observing double stars themselves, but if you are interested in observing planets, by measuring the closest double star observable this will really get a true measure of your scopes capabilities. The closer the double star you can see, the better the detail you should be able to see on a planet's disk. You might just be surprised to find out how good your setup really is. On the other hand, it might really show you how bad it really is and scream at you to replace it (or the eyepieces) with something more suited to what you enjoy looking at and how you observe.

3.2 A Long Hard Look at Your Equipment

Let us now take a look at the equipment you already have and why you originally purchased it. This chapter will also give you a few pointers to think about if you are thinking of purchasing a telescope, or are thinking of upgrading.

When your interest in astronomy developed, you inevitably purchased glossy magazines from the news-stands. A quick flick through them reveals pages and pages packed full of adverts of telescopes and equipment of all different sizes, shapes and to suit all budgets. When a person starts a hobby they really do not know if that interest is going to last. As a result their first acquisitions are usually fairly restrained, keeping the budget reasonably low. They will have little experience about what makes a good telescope for their kind of observing, that's if they even know what area of observing they are truly interested in.

With regards to accessories it is quite often difficult to know which ones would really enhance their observing experience until they pack a great deal of observing experience behind them. Much of the time the optical performance of many beginner telescopes do leave a little to be desired, especially if they were purchased in high-street outlets.

First and foremost, if you are thinking of buying any astronomical equipment it is always best to go to a specialist astronomical retailer. The proprietor should know a thing or two about astronomy, being able to offer you impartial advice. In the course of your discussions they will ask you relevant questions to gather enough information about your interests. They should then be able to identify just what type of scope is best suited to your needs.

3.3 General Notes About Telescopes

The main purpose of a telescope is to collect light not, as most people believe, to magnify the image. It does follow that the more light you collect, the more you will be able to magnify the image. The bigger the aperture of the telescope the more light it will be able to collect and the fainter the objects that can be seen through it.

If you are more interested in looking at the planets or the Moon, then a refractor is usually the recommended telescope of choice. A refractor produces less internal reflections, so produces a crisper image and is more suited for higher magnifications. If you can afford it you might extend your budget to buy a apochromatic refractor. This uses a design of lens that has a high reduction of both chromatic and spherical aberration, giving even cleaner and crisper images. Refractors, because of their multiple lens components are much more expensive per inch than a reflecting telescope.

If you like to observe faint clouds of gas, or distant galaxies, then a larger reflector, usually a Newtonian would probably be better. This has a larger aperture and collects a lot of light making faint objects look much brighter. The best thing about Newtonians is that they are cheaper to manufacture, so are much more affordable.

A 10″ reflector wouldn't require a second mortgage to acquire, especially if it is on a simple Dobsonian mounting. A Dobsonian mount is named after its designer John Dobson an American popularizer of astronomy. Due to its simple design it has bought down the cost of a reflecting telescope, especially large apertures, to a much more affordable level. If you were to look around for a quality 10″ refractor its cost would be literally astronomical.

This generalization sounds all well and good, but the telescope design types available today extends well outside this simple choice. Not only do we have "simple" refractors and reflectors, but also Cassegrain's, Schmidt-Cassegrain's, Ritchey-Chretien's, Schmidt-Newtonian's, Makzutov-Newtonian's and many, many more.

Another thing to take into consideration is the focal length of the telescope. A telescope with a longer focal length will generally produce a narrower field of view and a larger image, but the image will be fainter. It will be considered a "slow" f-ratio optical design. A short focal length telescope will have a much wider field of view, with a brighter but smaller image, so will be deemed a "fast" instrument.

So given this wide range of telescopes to select from, how do you know which type of telescope would be right for you? The only real way to answer this is to identify what it is you want to observe and how you do observe.

One more list to do before we move on.

What telescope did you buy first once your interest urged you to get one?
Was it a refractor or a reflector? Why did you choose this?
Did you opt for a big telescope or small for portability?
Did it have an alt-azimuth or equatorial mount?
Was it computer controlled with full GOTO or was it manually controlled?
What must-have accessories did you buy to go with it?
How much of your equipment gets used regularly?
What do you feel lets your equipment down?

3.4 Binoculars

These are probably the best value for money instruments that the observer can buy. Even an advanced observer still likes to keep a pair of binoculars handy, especially when scanning low to the horizon to find an object almost hidden in the twilight. They have low magnification and a wide field of view. (The author is not going to enter into the semantics of whether these should be termed binocular or not. As binoculars is the most common appellation, this is the term used throughout the book). Binoculars come in various varieties, with their optical arrangement normally given in the following format: 8×40, or 10×50.

The first figure is the magnification of the image produced. The higher the magnification, the more detail you will see, but the fainter the image will be. Also at a higher magnification the binoculars are hard to keep steady when hand-held. Anything above $12 \times$ magnification isn't really recommended unless firmly mounted.

The second figure is the diameter of the objective lenses in millimeters. The wider the diameter, the more light is collected, the brighter the image and the fainter objects that can be seen. As with most things, there is always a drawback. The bigger the binoculars become, the heavier they are and they become much harder to keep steady to obtain a clear view, unless they are mounted in some way.

Some of these stability problems with the larger binoculars can be overcome by attaching them to a tripod to keep them steady. A cheap adapter can be obtained from camera shops to achieve this way of improving the image obtained with them. Some people prefer to carry around relatively small and indiscrete 8 × 40 binoculars (probably the smallest recommended for astronomy), while other observers prefer larger binoculars like 10 × 50's to see fainter objects. But these are much bulkier and a bit harder to keep still while observing.

There will probably be another figure on the binoculars telling you the field of view in degrees. A typical 10 × 50 pair of binoculars typically has a field of about 5°. This is about the same distance between the two pointers, Dubhe and Merak in Ursa Major, or 10× the width of the Full Moon.

There are a number of image stabilized binoculars out on the market these days. Receiving favorable reviews, they are extremely good at keeping the image steady while observing, improving the view of the sky enormously. The biggest drawback is that they are still very expensive.

3.5   Choosing Binoculars

If you are thinking of getting binoculars, make sure that you try them out first to make sure that they are suitable for you before you purchase them. Look through them in daylight to check if the optics give any color fringing around objects viewed. Look at the edges in particular. If the image looks a little bit off or has colored fringes, they will look even worse in low light conditions looking at bright stars on a black background. Does the image reach a really sharp focus, or is it almost-there? Try looking at something outside against the bright sky with high contrast (Don't look through the shop window, stand outside if you can). Look at a straight object and move it from one side of the field of view to the other. There will inevitably be some distortion, but is there too much?

Whilst looking through the binoculars, close your eyes and open them again. If you quickly see the two images from each eye separately before your eye adjusts, this will lead to eye strain, so reject them. Hold the binoculars away from your eyes and look through them. Does the circle of light look round, or does it look diamond-shaped? If it looks diamond-shaped then the prisms are too small and are blocking light that will make the image darker.

The Mechanics

Does the focusing wheel operate smoothly, or does it catch or stop during the movement? Do both images from each eye register correctly, or do you feel yourself going cross-eyed while looking through them? Does the right eyepiece adjustment feel smooth, or does it catch or stop whilst turning? Does the central hinge or any of the moving parts move or rock when using light or moderate pressure? Is any

parts stick or feel loose reject the model immediately. What if you shake them? Can you hear anything rattling? If so, reject them.

Your Personal Comfort

This is highly important. If you don't feel comfortable using them, the chances are that you won't use them. Whilst testing for optical and mechanical quality, how do the binoculars feel when holding them? Do they feel comfortable to hold? Can you hold them up to your eyes for any length of time, or do they feel far too heavy? If they fail to meet any of these basic requirements, don't buy them. Keep looking until you find something that suits you properly, including the way you use them, meeting your budget as well as having as good optics as you can afford.

If you are satisfied that the binoculars are suitable, have another quick scan at them. Check the lenses, are they unscratched? Is there evidence of tampering?

Once you are satisfied with your binoculars, buy the pair you have tested. It's tempting to pick up the unopened box and purchase that one. Other binoculars on sale sealed in their boxes may not be as good as the ones you have already tested. Once you've found a good pair of binoculars it will be one of the best optical instrument that you will never regret buying and like most equipment, if looked after, will give you many years of service.

For a review of different types of binoculars see the Cloudy nights Web Site: http://www.cloudynights.com/ubbthreads/postlist.php/Cat/0/Board/binoculars

3.6 Telescopes

A number of major decisions need to be made and lots of questions need to be asked before you can even think of buying or upgrading your telescope. Many a beginner is amazed at the size of the box that arrives at their door when their new scope arrives. Budget and circumstances will dictate the type of scope you will buy and use regularly.

If your budget is a bit more substantial and you can leave everything set up between observing sessions, or storage space isn't too much of a problem, then you will be able to purchase a much bigger scope.

Bear in mind where you live. If you live in an urban area with an reasonably high level of light pollution, then a larger reflector would not perform as well as you might expect. The bright sky background would be "brightened" by the collected background light pollution. When observing many deep-sky objects there will be little or no contrast. As a result these faint objects will tend to fade into the background sky-glow and be extremely hard to make out.

Is storage room a problem? Would you want to take it out to darker skies to get the best observing conditions? In this case you might want to consider getting something a little smaller that is much more portable, easier to move about and quick to set up. A small portable grab-and-go setup is preferred by many observers these days.

Would you need to get it out of storage and set it up every time? If stored inside the house, once set up, the telescope will need time to cool down for the temperature to equilibrate with the outside temperature. This will allow convection currents to reduce enough to give better views. This is likely to take an hour or more. If you have an equatorial mount that has to be taken outside and set up every time, it would need to be polar aligned every time to be of maximum benefit. This setup time eats into your valuable observing time and can add frustration and annoyance to the session, especially if your choice of setup isn't quite right at the start. This will definitely take the edge off the enjoyment that any following observing session would undoubtedly bring.

The minimum that has to be done before you get your eye on the eyepiece, the more likely you are to get out and observe. If things become a real drag every time you want to observe, the less likely you are to get out and do anything at all. If it does take some time to set up, the chances are that the clouds will roll in and spoil things just as soon as you get started. Unless you are very dedicated, these niggling annoyances will soon dampen your spirit and you will find that you do less and less observing. Once this happens the more likely you are to feel that the outlay on equipment has been a waste of time and money as it just isn't getting used enough. Under these circumstances you can very quickly begin to lose interest and give up the hobby for good.

To identify what sort of telescope will work for you, get your eye on the eyepiece end of as many telescopes as possible before you commit your hard earned money. Get together with your local astronomy club. These quite often hold evenings where a number of members take their telescopes out observing under a dark sky. Grab a peek through as many of their scopes as you can. Only by experiencing what the views are like through a particular type of telescope (and eyepiece) can you see if it could be the scope for you. Can it be set up and running very quickly? Are you excited by what you can see?

Cost will obviously be a big decider in what you buy as well. One word of advice is to buy something that you really do have to stretch a little bit more to afford. If you need to wait a little longer to save up for it, then do that. The wait will be well worth it in the long run as you will get a telescope that has greater potential for later development. For those who have decided to take the plunge and buy a scope or upgrade, the big question now is, do you buy new equipment, or second-hand?

Buying new equipment from an reputable supplier gives somewhere for you to go back to if you do have subsequent problems. This gives complete peace of mind. They will also be able to offer a lot of advice on other accessories after you have made your purchase. If you have a local supplier, pop in and have a chat with them. They are usually very friendly and will discuss your particular requirements and match those to a scope that more suited to your needs. They will often also give you advice on what to observe or where to find out more information.

Buying second-hand equipment can save you a lot of money. Some observers regularly change or upgrade their scopes and accessories. This ensures that there is a constant stream of good quality second-hand gear regularly coming up for grabs at bargain prices. These frequently turn up on For Sale notice boards on astronomy forums or on astronomy second-hand Web sites. Some suppliers also advertise second-hand equipment on their Web pages. You might also occasionally see second-hand scopes appearing in adverts in your local newspaper or online general for-sale Web sites. So keep a look out and you might just find the scope you've set your heart on soon winging your way at a bargain price.

If you do decide to look at the second-hand market, you really do need to know what you are looking at before you buy to make sure that you get the bargain you expect. If in doubt, ask someone you know who has a bit more experience. The important thing to remember is: Caveat emptor. Let the buyer beware! Always be careful sending cash to a stranger.

Having quite a bit of experience at looking at all sorts of equipment that is available, before you take the plunge, will help you identify good offers when they do present themselves. By now your developed keen eye should help you to tell on a quick inspection whether the equipment advertised has been treated well and will perform as expected once you get it home. If you can use it under a clear sky before buying it, all the better. Many observers use scopes purchased second-hand. If you do buy something and later find that it doesn't quite suit you, it can always be sold on again. The second-hand market is quite good in the fact that if the equipment is still in very good condition when it hits the market again, it can usually be sold for about the same as the second-hand price you originally bought it. Of course those with more advanced DIY skills will be able to construct telescopes themselves.

If you stick with the hobby, you will almost without question find yourself gradually making your way through a number of different setups before you do settle down on something that really works for you. Hence adding your own used equipment onto the second-hand market for other observers to snap up as bargains as they too develop.

Chapter 4

Accessories

4.1 Finder Scopes

To get objects centered in your field of view you do need to make sure that you have a good finder scope on your telescope and that it is correctly aligned with the main scope. This will enable you to find the object you are hunting for (or a bright object close to it) in the wide field of view. If aligned correctly, you should then be able to look through the main telescope and see the object in the higher magnified and much narrower field of the scope. A variety of finder scopes available these days in order to achieve this. All types have their advantages and disadvantages.

A Traditional Finder Scope

This is essentially a small telescope mounted onto the main telescope. They frequently have a similar light grasp and magnification of one half of a pair of binoculars, something around 10×50. When you look through the smaller scope, having a shorter focal length and a low power eyepiece it has a wider field of view than the main scope. It usually also has a cross hair built into it which is at the focal point of the eyepiece, so easily visible when looking through it. These finder scopes come in two main types, a straight-through system where you look straight through the instrument, and right-angle finders where the light path is bent at right angles. The latter certainly means that the eyepiece of the finder is often in a more accessible position, but some observers just do not get on with them as the change in orientation of the field of view can be disorientating.

You need to ensure that the wider angled finder is adjusted and aligned properly with the main scope. Failure to do this makes the finder virtually useless. The finder

D. Eagle, *From Casual Stargazer to Amateur Astronomer: How to Advance to the Next Level*, The Patrick Moore Practical Astronomy Series, DOI 10.1007/978-1-4614-8766-1_4, © Springer Science+Business Media New York 2014

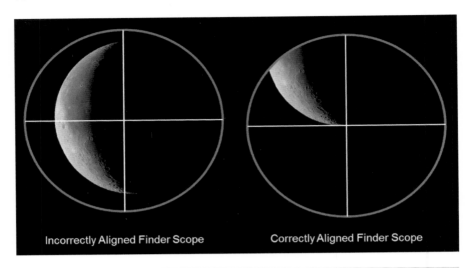

Incorrectly Aligned Finder Scope   Correctly Aligned Finder Scope

Fig. 4.1 View through the finder scope and aligning with the main telescope (Courtesy of the author)

can be adjusted correctly by aligning the main telescope on a distant terrestrial object or the Moon. Make sure that the object you have chosen is in the field of view and a particularly distinctive part of the object is central in the field. Now have a look through the finder and see if it is pointing at the same object. If the finder scope is incorrectly adjusted, the object may be to one side of the finders field of view, or if really bad, not in the field of view at all. There are usually three adjustment screws that can be used to move the finder within its mounting. Use these screws in turn to move the finder so that the object you have chosen is also aligned correctly to match what you saw previously in the main scope.

It may take a few attempts to get it spot-on, but it is always well worth the effort. Don't forget that if you are using an un-driven telescope you will need to keep re-centering the alignment object in the main scope regularly, unless you are in the northern hemisphere when you will be able to use Polaris as your alignment object.

Once aligned properly the two scopes will complement one another well, making sure that you are able to place the position of your object of choice on the cross hairs of the finder first, before moving over to the main telescope where the object should now be in the field of view. Having a correctly aligned finder scope will make targeting objects much easier. Try re-centering the finder on a different object and see how close the object is in your main telescopes field of view. If everything is set up correctly, you should be aligned on each object every time.

Red Dot Finder

These types of finders are becoming very popular. These are battery operated and when the observers head is held in the correct position, a red dot (or in some cases an illuminated target) is superimposed onto the sky. The position of the telescope is moved until the dot or target is superimposed onto the right part of the sky. A popular choice is the Telrad. One disadvantage of these types of finders is that they are exposed to the elements when in use. In damp conditions they quite often dew up throughout the night and have to be dried before they can be used again.

Laser Finder

These finders project a bright laser beam into the sky. The resulting beam can be seen to project into the sky "pointing" up into the sky. You move the telescope until the beam is pointing at the point of interest and the object should be in the field of view. Make sure that if you do buy a laser finder, that you buy one that is legal for your country. There are a number of extremely bright lasers that are available on the Internet. These should be avoided at all costs.

There are two major downsides to these types of finders:

1. When being used they produce a beam that would be visible to people passing by. This might attract some unwanted attention.
2. You must be very careful if aircraft pass overhead fairly low as they could be interpreted as someone trying to dazzle pilots. There have been known prosecutions of people using them inappropriately, so make sure you do not inadvertently cause problems. People have been known to be prosecuted for causing a nuisance with these laser pointers, so please take care if you do use one.

The last few types of finders take batteries. Although they cost about the same as a 10×50 finder, there are some associated running costs. They are also another thing you can forget to turn off at the end of a long tiring observing session. This can resulting in you coming back for your next observing session only to find flat batteries. So always keep some spare batteries handy.

Mobile Phone/Electronic Device

There are a number of electronic devices that can be used to point the telescope in the correct position. These range from cheap Apps that can be used on a mobile phone to dedicated digital angle gauges. These will need to be calibrated with the actual sky in order to work correctly. Some observers reportedly use them regularly, but others find them a little cumbersome to use.

4.2 Eyepieces

An eyepiece can really make or break a telescope. Use a poor quality eyepiece on a really good telescope and you will never get more than a mediocre or poor image whilst looking through it. The eyepieces that are often supplied with some of the budget telescopes do in many cases do leave a lot to be desired. Many scopes like this really under-perform because of the inferior eyepieces supplied with them.

A really good investment to the hobby is a selection of eyepieces that will complement the optics of your telescope. There is no sense in paying a lot of money for good telescope optics if the eyepieces used with it drag down the quality of the image.

Like the telescopes you use them on, eyepieces come in many guises. The higher end of the budget you can afford, usually the better. Try and get a look through someone else's eyepiece, or even use it on your own scope to see how it performs, before you buy. Good advice would be to buy the most expensive eyepieces you can afford within your budget. Of course, as each eyepiece can be purchased individually you don't need to splash out and buy all your eyepieces at once. This additional expense can be spread out over many years. Once you have purchased your collection of eyepieces, if properly looked after, they can last a lifetime. This means that as your observing "career" develops and you do eventually update your telescope (this is inevitable if you do stick with the hobby), you will keep hold of your eyepiece investment/s and use them on your later scopes.

Properties of an Eyepiece

An eyepiece usually has a number of values which determine what you will see:

Focal Length
Usually expressed in millimeters.

This is the measurement that dictates the magnification you will obtain when used on any particular telescope.

To calculate the magnification of the eyepiece on your telescope, divide the focal length of the telescope (in mm) by the focal length of the eyepiece being used.

E.g. a telescope of 1,000 mm focal length used with an eyepiece of 20 mm focal length gives a magnification of 50×.

1,000/20 = 50.

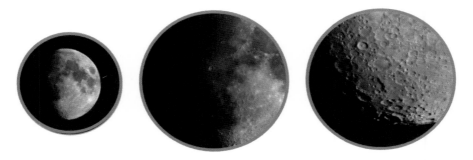

Fig. 4.2 Comparing eyepiece field of view at different magnifications (Courtesy of the author)

Fig. 4.3 Comparison of wide angle and normal eyepiece field of view (Courtesy of the author)

Due to the instability of the Earth's atmosphere and unsteady seeing, the maximum usable magnification is usually not much more than 300x. You don't really want to get an eyepiece that will bring your magnification above this figure. There will be some exceptional nights when this magnification may be exceeded. Sadly these observing conditions occur very rarely.

Field of View

This is the area of sky that is seen when you look through the eyepiece. This can easily be calculated by letting a star drift in from one side of the field of view and exit the other side. The time measured in seconds gives the field of view in arc seconds ("). This will depend on a variety of factors:

1. The focal length of the telescope being used.
2. The focal length of the eyepiece.
3. The design of the eyepiece.

Some older eyepieces have a very narrow field of view. This makes the view very much like looking down into a tunnel. Eyepieces which have a wider field of view (but the same focal length and therefore magnification) give wonderful views.

You will pay a lot more for the privilege, especially for well corrected eyepieces where pin sharp stars are visible right to the edges of the field of view.

Eye Relief

This is the measure of optimum distance that the eye needs to be behind the eyepiece to get the best view. Usual range is between 5 and 20 mm. The closer to the eyepiece, the more problems you have with eyelashes showing up in your view.

Fig. 4.4 A range of eyepieces for different objects. Long focal length = Wide field for deep-sky. Short focal length = Planets (Courtesy of the author)

Eye relief isn't usually much of a problem for most people, but if you wear spectacles this will become very important. In this case the spectacle wearer will need to have their eye further from the eyepiece to accommodate the glasses. Someone in this situation will need a eyepieces with longer eye relief. A disadvantage of a longer eye relief is that the further from the eyepiece the observers eye is, the harder it is for the observer to keep their head still in the correct position for the best view.

Exit Pupil

This is the diameter of the light path as it exits the eyepiece on the telescope, usually measured in mm. If this is wider than the widest diameter of your pupil when it is fully open, then some of that light will not enter your eye and is effectively wasted. This means that your telescope is not operating at optimum and objects will look fainter than they ought to be for the scopes light collecting area.

As you start to collect eyepieces, it is always best to get them focal lengths that are not multiples of one another. You can also purchase eyepieces that are Parfocal. These, when swapped from one eyepiece to another, do not have to be re-focused each time an eyepiece is changed. It will save you having to refocus every time you change eyepiece. A simple Parfocal ring can be obtained to fix to each eyepiece barrel to achieve the same thing, but much more cheaply.

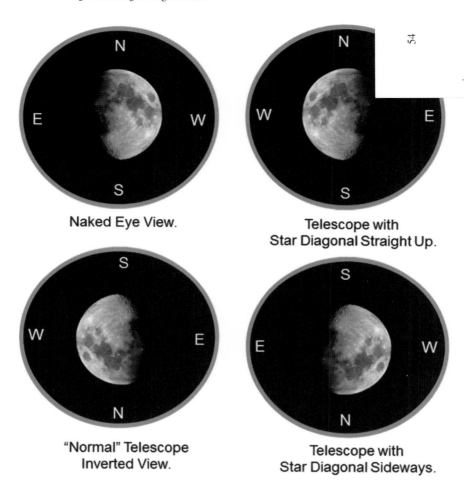

Naked Eye View.

Telescope with
Star Diagonal Straight Up.

"Normal" Telescope
Inverted View.

Telescope with
Star Diagonal Sideways.

Fig. 4.5 Orientation of field of view changes when right angle prism is used (Courtesy of the author)

4.3 Star Diagonal or Right-Angle Prism

These are used on refractors and make it easier for the observer to view the image. But observer beware! The orientation of the field of view changes dramatically, depending on which way round and the angle at which the star diagonal is inserted into the telescope. Always re-orientate yourself so you know the compass points in your field of view.

Fig. 4.6 Right angled Prism in use on a small telescope (Courtesy of the author)

4.4 Barlow Lenses

Hand in hand with eyepieces come Barlow lenses. These optical components are slotted in between the telescope and the eyepiece. They increase the focal length of the telescope. They come in different powers. A 2× Barlow lens will effectively double the focal length of a telescope. So a 1,000 mm focal length telescope will become a 2,000 mm telescope. They also come in 3× 4× or 5× powers. When a 2× Barlow lens is fitted into your telescope, your 20 mm eyepiece will now have the same focal length (and magnification) as a 10 mm eyepiece without the Barlow.

Don't forget that the increased magnification will do a number of things: Reduce the field of view, magnify the image and decrease the brightness of the image. It will also magnify the damaging effect of atmospheric interference.

In our example above, we already have an eyepiece of 20 mm length and fitting a 2× Barlow gives us the equivalent of a 10 mm eyepiece. Having a single Barlow lens will instantly double the number of possible magnifications available from your existing eyepieces. So if you already have a 20 mm eyepiece and have a 2× Barlow (or intend to get one later), investing in a 10 mm eyepiece won't give you any gain (but a lot to lose from your wallet). In this case a 15 mm eyepiece would be a good investment, giving the equivalent of a 7.5 mm eyepiece when used with the Barlow. So choose the focal length of your eyepiece collection wisely. This will you a much wider range of possible magnifications available for observing different objects, without the cost of buying too many expensive eyepieces.

Quality of the Barlow is however quite important. If you have invested in a good quality telescope, and eyepieces, then make sure that the quality of the Barlow lens is up to standard. Otherwise you will not get the best out of your telescope when using the Barlow. A Powermate is also a type of Barlow lens manufactured by Televue. The optical arrangement is different, but they are top quality, optically corrected to give superb images. But they do come at a cost.

Use David Paul Green's online Eyepiece Calculator to determine the magnification of eyepieces when used on your telescope (with and without Barlow lenses): http://www.davidpaulgreen.com/tec.html

4.5   Filters

Using filters for observing on your current set-up will help to enhance what you can see. There are a wide number of filters now available for astronomical use.

Filters come mounted in a metal ring with a thread which screws into the barrel on the end of an eyepiece. They come in 1.5, or 2.0 in. sizes depending on the size of the eyepiece on which they will be used. Filters that fit two inch eyepieces are bigger and therefore much more expensive.

Filters work by allowing certain wavelengths of light to pass through, and block out others. There are a number of deep-sky filters available but their use is fairly limited especially if used in a bright urban area. There really is no substitute for observing in dark skies. For someone interested in planetary nebula an OIII filter will allow the light of the planetary nebula through, but most of the light from the surrounding stars will be mostly lost. When the filter is in place the stars fade, but the planetary nebula stays bright. By moving the filter between the eye and the eyepiece the steady planetary can be identified, as it shines with a steady light and the stars appear to blink as the filter intervenes.

Colored Filters

The appearance of features on the planets can be also be enhanced by using different colored filters.

As well as the color that the filter transmits many of them are assigned a number, referred to as The Wratten Number. This system has been in use since 1912 when Kodak started producing colored filters for photography.

Filters vary widely not only in the wavelength they transit, but also in the range of wavelengths. Some let light through in a very narrow range of wavelengths, others are less restrictive, letting a broad range of wavelengths through. The wavelengths that they do transit will determine their usefulness on different objects. These are used to enhance certain features on an astronomical object. You use a color of filter opposed to the color of the feature that you would like to observe.

At some time or another you may have come across the idea of color wheels that inform us which colors complement one another. The color wheel is useful in trying to determine which color filter to use to enhance a particular object.

For an example of a color wheel see this Web link: http://www.tigercolor.com/color-lab/color-theory/color-theory-intro.htm

If we are looking at, say Jupiter, and want to enhance the red spot a bit better, we can look at the color wheel and can decide what filter would do the job for us. As the spot is indeed red (but sadly not as red as it used to be) look on the opposite side of the wheel to see its opposing color. In this case green. Indeed to enhance the red spot, or any other red feature, a green (Wratten 58) filter is used. This same filter is also used to accentuate lunar surface contrast and melt lines around the Martian polar caps. A blue filter (Wratten 38) would be used to enhance Martian clouds, and a deep blue filter (Wratten 46 or 47) can be used to enhance faint cloud details on Venus.

For details of the effects of filters on observing the planets from the Association of Lunar and Planetary Observers visit: http://alpo-astronomy.org/mars/articles/FILTERS1.HTM

Nebula Filters

There are also a number of so-called nebula filters that are designed to increase the chances of seeing certain objects. They work by excluding certain wavelengths of light and excluding those that are not emitted by the nebulae. This results in an increase in contrast between the nebula and the background sky.

There are many different designs, but the main ones for visual use are:

Deep Sky and UHC Filters

These are broadband filters which block the transmission of the coming from high and low pressure sodium lamps, neon lights and airglow. This improves contrast between the nebulae and the sky background.

Hydrogen Beta Filter

Narrow band filter in the red end of the spectrum. Useful for revealing the faint nebula that the Horsehead Nebula intrudes into and the Cocoon Nebula in Cygnus.

Oxygen III Filter

A narrow band filter down the blue-green end of the spectrum. Can enhance the appearance of planetary nebula. Also works extremely well on revealing the Veil Nebula in Cygnus.

Many observers report that the cost of the filter may not warrant the small improvement in view it brings. The performance of a filter will vary in each telescope setup, observing conditions and weather. Every observer will of course also have their own opinion.

Light Pollution Reduction (LPR) Filters

There are filters available that block certain parts of the spectrum and let through most of the other light you want. These are especially useful as Light Pollution Filters. These let most wavelengths of light though, but block the blue and yellow part of the spectrum where most sodium lights transit, letting red and green light through. This should produce a darker background sky as most of the interfering light pollution is removed. The red and blue light is most prevalent in deep sky objects like nebulae, making the object look a little brighter. This type of filter will improve the contrast between the image you want to observe and the background sky, making the object easier to see.

There are a number of light pollution filters available today. The price range of these filters ranges enormously and each has its own distinct properties. What works best for you will have to be found by trial and error. If you can find someone who has a filter why not ask if they will let you try it out on your equipment before you buy one.

Deep Sky & UHC Filter

Best suited to big telescope that collect a lot of light.

Orion Skyglow Filter

A general good budget priced filter.

Baader Neodymium Filter

These were originally designed to accentuate features on Mars, Jupiter and the Moon. It was soon found they were also quite effective at cutting out light pollution as well. These dual purpose filters are very reasonably priced.

Skywatcher Light Pollution Filter

About half the price of the Baader Neodymium filter, this budget LPR filter does what it says on the tin. But at this price has some limitations.

Astronomik CLS Light Pollution Filter

About the same price as the Neodymium filter, it can also be used for DSLR imaging. Also acts as an effective minus violet filter for imaging on small refractors.

HUTEC IDAS Filter

Mostly used for imaging, so will not usually be of much benefit to the visual observer.

ALWAYS REMEMBER there is no substitute for observing in a dark sky. As well as filtering out the unwanted light, light pollution filters will also decrease the amount of wanted light that will reach the eye. The resulting image will be a bit darker as a consequence. In many cases the filter also produces some unwanted color casts to the viewed image. Filters are also used extensively in imaging, but this is not going to be covered by the scope of this book.

For a description of how different filters are used in astronomy:

http://sas-sky.org/wp-content/uploads/2011/09/SAS-The-Use-of-Astronomical-Filters1.pdf

For a primer on Wratten Numbers: http://en.wikipedia.org/wiki/Wratten_number

For a list of a wide range of different filters used for astronomy, and their comparisons, as written by both the Sangamon Astronomical Society, the Prairie Astronomy Club and the Cloudy Nights Web Pages, go to:

http://sas-sky.org/wp-content/uploads/2011/09/SAS-The-Use-of-Astronomical-Filters1.pdf

http://www.prairieastronomyclub.org/resources/by-dave-knisely/filter-performance-comparisons-for-some-common-nebulae/

http://www.cloudynights.com/item.php?item_id=63

http://www.cloudynights.com/item.php?item_id=1520

4.6 An Occulting Bar

When looking for faint planets, such as Phobos & Deimos close to the bright glare of Mars, or when trying to view a much fainter companion to a much brighter star, an occulting bar within the eyepiece can be a very useful edition to your armory.

Exercise 4A: Make Yourself an Occulting Bar

These are not available commercially but they can be made quite easily from a roll of thick black card and black electrical insulation tape. Cut out a rectangle of black card about 1 in. by 4 in.. Carefully wrap the card around itself and fit it inside the barrel of one of your existing eyepieces. Leave one side of the card exposed and wrap insulation tape around that end to keep the card in place. Remove the loop of card from the barrel and turn it around and place back into the eyepiece barrel. Secure this side of the card as well. Once the card loop has been secured, remove the card from the barrel. Make sure that the loop fits in the eyepiece barrel quite snugly so that it does not fall out very easily. Wrap some more black tape around the roll of card to make it fit better if necessary. Place the loop of card on top of another piece of black card and draw around the ring. Cut out this circle of card and then divide it by cutting directly across the middle of the circle to form a semi-circle. Secure this new piece to one end of the ring of card already formed using black tape.

The resulting occulting bar can now be used on most of your eyepieces to blot out bright objects in the field of view in order to see fainter objects close by. Make sure that you insert the occulting bar with the occluding piece first so that it becomes as close to the focal point of the eyepiece as possible. Rotating the eyepiece with the occulting bar in place will enable you to view different parts of the field of view by changing the area of sky blocked by the occulting bar.

For more detailed instructions on how to make a simple occulting bar for your eyepieces, download the instructions from the author's Web page:

www.eagleseye.me.uk/OccultingBar.html

4.7 Dew Prevention

There you are happily observing away and over the course of the night you will gradually notice halo's appearing around bright stars. Slowly but surely the fainter stars are seen to gradually disappear. A glance upward confirms that the sky is still very much clear of clouds. You can always tell when a telescope has reached the same temperature as the surrounding air as the objective lens, corrector plate or mirror will often mist up. This will also affect eyepieces left out in the open. The problem is usually worse with refractors or reflectors that have a lens or corrector plate at the end of the tube closest to the sky. As the telescope temperature gets below the dew point, condensation will inevitably form on the surface. It is best not to wipe this off! Wiping the optics can scratch a lens or damage the mirror's surface. The easiest and best way to remove the condensation is to use a small hairdryer (Careful with electricity in damp weather). Simply use the lowest setting and blow warm air over the optical surface. The heat will evaporate the moisture droplets clinging to the optical surfaces bringing back your crystal clear images. Don't make the mirror or lens too hot as this could distort the optics, causing undue image. Condensation of water on the objective or surface of the corrector plate can be prevented by adding a dew shield to the end of the telescope tube. This extends the tube so that the coldest of the outside air is kept away from the optics for as long as possible.

This method only delays the time that the optics reach dew point, and does not prevent it dewing up altogether. The most effective way of preventing the optics from dewing over is to use a dew heater. This is a high resistance wire that wraps around the objective lens, corrector plate or eyepiece and keeps it slightly warm.

Fig. 4.7 A home-made dew cap fitted on the authors 190mm Mac-Newt (Courtesy of the author)

ly designed to operate using a low voltage so they are extremely
.e open air in the damp.

lave finished observing and you have to bring your telescope and
k inside, you may find that they become coated in dew as soon as
n in from the cold. This will not hurt the telescope and optics, but it
.ouch the telescope. If the lens or corrector plate is misted over, leave
everything that has dew on it open. Leave the lens cap off, if you didn't replace it
before bringing the scope back in. Leaving it for few hours, or overnight at room
temperature, the condensation will have evaporated off. Once dry place all the caps
on to prevent dust building up and then store in a dry place that has as little dust as
possible.

4.8 Collimating Your Telescope

One crucial thing that many observers, even very experienced veterans, often fail
to get right is the collimation of their telescope. This is the alignment of the mirrors
and optics within the focal path of the telescope. Correct alignment is crucial for
you to get the very best image out of your instrument. It really does have to be
spot-on.

Collimation is the process of getting the optical components correctly aligned in
the telescope so that the light path is as straight as possible. Only when a telescopes
optics are properly aligned will it will give its sharpest possible image.

Many observers have a real problem committing themselves to doing this essential task. Most deem it too difficult to perform themselves, especially when first
starting out. But it cannot be stated enough that it is imperative that it is done.
So whatever you do, don't shy away from the task. If you get the collimation right
you will notice a big difference in the quality of the image produced by your equipment. If the Airy disk produced by your telescope looks like Fig. 6.5, then your
telescope definitely requires re-collimating.

Most refractors should not need collimating as their lenses are usually fixed into
place. If the collimation of a refractor is out, then the telescope has probably been
dropped in the past, or is of such poor quality that you really shouldn't be using it
at all for serious astronomical observations.

Reflectors will need regular checking of their collimating, especially if they are
moved around frequently. Make sure that you take the time, whenever you
have the opportunity, to check regularly the collimation of your scope. If you do find
the collimation is out, take time to learn how to adjust it properly.

Reasonably good collimation can be achieved by eye, especially when well-
practiced. It's all to do with getting all the circular edges you can see in the field of
view as perfectly concentric as possible. Laser collimators can be purchased which
can be used as an aid to collimating and these are very useful for collimating
Newtonian telescopes. You must be very careful if you do use a laser collimator as
serious eye damage could result if used incorrectly.

Sky & Telescope have a fine tutorial on collimating a Newtonian telescope:
http://www.skyandtelescope.com/howto/diy/3306876.html
Astro-baby also has a guide to collimating a Newtonian telescope:
http://www.astro-baby.com/collimation/astro%20babys%20collimation%20guide.htm

Schmidt-Cassegrain and Maksutov-Newtonian telescopes are a little bit more difficult to collimate than simple Newtonian reflectors, but these guides should help you.

Ed's guide to SCT Collimation on Astromart:
http://www.astromart.com/articles/article.asp?article_id=548
Collimation of a Maksutov-Newtonian on Cloudy Nights:
http://www.cloudynights.com/item.php?item_id=396

Once you are confident that your telescope is collimated correctly, the real test of your telescopes performance and the optical alignment is to look at a stars image and see what it looks like.

4.9 Cleaning/Resurfacing Lenses and Mirrors

Best word of advice is not to touch the mirror at all. Most mirrors supplied in reflectors these days are finished with a transparent coating which protects the silver mirror surface. The aluminized surface of modern mirrors is quite resilient these days and need not require a re-coating for a good number of years if properly looked after. A slight coating of dust on the mirror will not affect the image too much. It always looks a lot worse than it really is. If you do decide to clean a mirror and you smear it during the process that will have a more detrimental effect on the quality of the image than a thin layer of dust. So in many cases, leave the mirror just as it is. In some cases, usually due to incorrect storage, the mirror could be so badly covered in dust a grime, it really will need to be cleaned. But always exercise extreme caution when doing do. If you do decide that the mirror really does need cleaning the best way is to do it very gently using extremely diluted gentle detergents.

If you feel that your mirror really does need cleaning, follow Astro-Baby's guide:
http://www.astro-baby.com/TAL%20Telescope%20Rebuild/Telescope%20Mirror%20Cleaning.htm
Sky & Telescope have a Caring for Your Optics Web site, which also includes details on how to clean a mirror:
http://www.skyandtelescope.com/howto/diy/3437191.html?page=1&c=y

Despite the modern coatings that modern mirrors there may come a time where the old coating will need to be removed and a fresh coating applied. It is always best to take the mirror to a reputable astronomical supplier for cleaning and re-aluminizing.

will find yourself equipped with your ideal setup. Unfortunately it doe. ᴜite stop there. The keen observer can always be tempted to acquire, say, a long focal length planetary scope to complement their large aperture deep-sky light bucket, or a small portable refractor for more mobile observing out in darker skies.

As you start seeing fainter and fainter objects, and want to see objects in greater detail, the temptation is to purchase a much bigger telescope with a larger light collecting area in order to see fainter objects. There never will be "One Perfect Telescope". Like most hobbies, astronomy can be as cheap or as expensive as the amount of money you want to spend.

Aperture Comparisons

The amount of light gathered by any particular instrument is dependent on the aperture and therefore the light collecting area. The light-collecting power of a telescope can be calculated by using good old high school math.

The area of a Circle $= p \times r^2$. So if we know the diameter of the mirror or lens we divide that by two to get the radius. We will take p as being equal to 3.142. So a mirror or lens with a diameter of around 3 in. (80 mm) will have a radius of 40 mm.

Area $= 3.142 \times 40^2$

Area $= 3.142 \times 1,600$.

Area $= 5,027$ mm^2

Let us perform this same calculation on a 6 in. telescope (160 mm) lens or mirror, being double the size of the 80 mm telescope calculated above.

Area $= 3.142 \times 80^2$

Area $= 3.142 \times 6,400$.

Area $= 20,108$ mm^2

So although it only has double the aperture of the previous telescope, it has over four times the surface area and light collecting power. Any viewed object will be about four times brighter when viewed through it. Let us now compare to a 10 in. (254 mm) telescope.

Area $= 3.142 \times 127^2$.

Area $= 3.142 \times 16,129$.

Area $= 50,677$ mm^2

This telescope has 2.5 times the light-collecting power of the 6 in. telescope and a massive ten times that of the 3 in. telescope.

The Faintest Stars Visible in Different Sized Telescopes

It follows that a telescope with a larger objective or mirror that collects more light will be able to resolve fainter objects. There are many more fainter stars than there are brighter ones. So the fainter you can see, the more interesting objects come into view (Table 4.1).

Table 4.1 Average Limiting magnitude achieved with different aperture telescopes

Size (in.)	4	8	12	14	20	20
Magnitude	12	13	14	14.5	15.3	16.2

This table is only a rough guide, but no wonder aperture rules in observing. But do try and resist catching aperture fever on the basis of these calculations. The type of telescope you have and your observing conditions will drastically affect how well your telescope actually performs, or not! The only way to determine the faintest star that you can see with your telescope is to get out there and look for yourself.

If you are thinking of getting a much bigger "light bucket" to see faint deep sky objects, you really do need to think about where you are going to do most of your observing. Think about where you live. If you have a reasonable amount of light pollution around you, a large aperture telescope will not perform as well as you might hope. As well as collecting light from your intended faint targets, the huge surface area of the mirror will also collect light pollution from the sky surrounding the object. This will result in sky-glow being visible in the image, reducing the contrast between the object and the background sky, making it much harder to spot.

4.11 Your Own Observatory

As your hobby develops and you really do enjoy observing, very soon your thoughts will inevitable turn to getting yourself some type of permanent observatory. This is the undoubtedly the best investment you will ever make to encourage your observing. But why is an observatory so beneficial? If a telescope is set up permanently it doesn't need to be set up each and every time the skies clear. The time you cut down on all that time spent setting up polar aligning, running cables, etc., can more effectively be used actually observing. You will also be sheltered out of the worse of the wind. You will be able to quickly go out and snatch a couple of hours if a short clear spell is predicted. If you need to set everything up every time you want to observe it soon becomes tiring and a bit of a drag, clouding over before you're ready, so many times you just won't bother. If your setup is made much quicker and easier, your time spent actually observing will increase dramatically. There are pros and cons of having an observatory.

On the plus side, you will be out of the wind and shielded from the brunt of the cold. Equipment is already set up and ready to go and you are hidden away from neighbors/passers-by.

On the negative side, it can be expensive, can produce some turbulence from the difference in temperature inside and out, and only a limited part of the sky visible at one time (if a dome).

As for what type of Observatory you should have, there are a huge variety of observatories available. They can be the stereotypical dome-shape, or a simple roll-off box built over your scope. They can be off-the-shelf items, custom built or even home-made. The observatory you will eventually acquire will depend entirely on a number of factors: Your budget, the space you have to house a potential observatory, your DIY skills and your partner's tolerance levels.

The siting of your observatory will depend on your personal circumstances. Many of us do not have very big back yards, so have to make do with what we have. Try and locate your observatory as far away from other buildings as possible. As well as restricting how low to the horizon you can see, these will also cause thermal currents, which will reduce the quality of the seeing. Try and ensure that you have a good southern horizon (or northern if you are in the southern hemisphere). The sky varies much more in these directions throughout the year. East and West horizons are also quite beneficial to be able to see the inner planets or a thin crescent Moon or the outer planets as they move away from or are approaching the Sun.

Also determine if you need planning permission to build your observatory. You may often hear people saying things like " It's not a permanent building", or "It's just like a shed", "so it won't need planning permission".

The only way to answer this question is to speak to the people who will be able to tell you with any authority if you do need planning permission. This will always be your local building planning authority. Book an appointment with them and speak to them in person, to discuss your plans with them. Take along pictures of your proposed construction with all the dimensions and the general location on your plot of land. You will still need to apply formally even if planning permission isn't required. Once your letter has been received by the planning office, they will look at the details. If planning permission isn't required they should send you a covering letter stating this fact. File this letter away safely as this will cover you for any problems, if a neighbor subsequently complains to them after the sudden appearance of a "strange" building in your back yard.

There are many reasons why planning permission might be required. Planning rules vary from region to region. It might even, in some cases, come down to the personal view of the planning officer within the planning office who makes the final decision. This is why trying to talk to them in person and getting a feel for their way of thinking is always a good idea.

The author originally got confirmation that planning permission wasn't required when he erected his observatory some years ago. When he moved home within the same town, he wrote to the same planning authority expecting to receive a similar letter stating he didn't need planning permission in the new property as before. He believed it would be exactly the same process with no planning permission required. How surprised was he to receive a letter stating that, in this house, despite it being exactly the same building, he did need planning permission. After speaking to the planning office it was discovered that this was due to some degree of public access round the back of the new property being so close to the re-location site of the observatory. There are many factors to be taken into account when building a construction like this and only your local planning department will have that complete information at hand.

So the moral of the story, never assume you don't need planning permission. Always make sure that you check and if you do get a letter stating permission isn't required, keep it safe.

Once you have established your own observatory you will never look back. As a result your observing productivity will increase enormously. You will be able to get everything up and running really quickly and take real advantage of short breaks in the clouds that you would otherwise miss. Your observing opportunities and ability will rise in leaps and bounds.

Heating in an observatory is a big no, no! The difference in heat between the warm inside and outside will produce turbulence. This will destroy the seeing and make everything look blurred with the image boiling away. To keep an observatory dry in wet weather the author has found it best to keep a couple of low wattage lights on inside. This helps to stop condensation forming inside without making the interior too hot.

When you want to use the observatory, open up the roof at least an hour before you want to use it. This will enable the temperature inside to equilibrate with the outside so that any thermal disruption to the seeing is minimized. This is particularly important at the end of a hot sunny day.

The following book contains many more ideas on the different types and designs of observatories that can be built:

Small Astronomical Observatories: Amateur and Professional Designs and Constructions. Patrick Moore (Ed.). Springer 1998.

The AllAboutAstro web site has some suggestions how to go about planning and building an observatory:

http://www.allaboutastro.com/Articlepages/observatoryonearticle.html

Sky & Telescope have a few suggestions on mistakes to avoid when building an observatory:

http://www.skyandtelescope.com/howto/diy/3305721.html

Before we move on, let us take a quick draw of breath at this point.

Does having the biggest, latest, best and most expensive equipment make you a good observer? The answer to this question is definitely no. Even an observer with the most modest of equipment is able to make very useful observations as long as he knows the limitations of his equipment and becomes very familiar with its use.

The only real way forward to develop your skills is to be properly prepared with planned and organized observing sessions. You should have a list of objects you want to observe and have searched for any new discoveries to make sure you don't miss these either. You should know the best conditions for hunting your chosen objects down and eliminate those objects you probably have no hope of seeing under current conditions. You should now have an observing log to record your observations. So let us have a look at what you might to observe and how to get the best information on each type of object to increase your chance of success.

Get to know your existing equipment and its limitations really well. Explore all the scopes capabilities and use those to the best of your ability. Once you are extremely comfortable and know your existing equipment inside out, will your observing really start to improve. The next step in your journey is to develop a greater understanding of when and how to observe different objects. So in the next few chapters we will be looking at some aspects of those celestial wonders that we can observe and the nature of their appearance.

Part II

Practical Observing: Your Quarry Awaits

Chapter 5

Constellations, Asterisms and Stars

Today there are 88 constellations officially recognized by the International Astronomical Union. Their official list is located here:

http://www.iau.org/public/constellations/

Virgo is the largest by area of sky it covers, Crux being the smallest. We all know Orion with its bright stars, distinctive belt of three stars in almost a straight line, ruddy Betelgeuse and white Rigel. But how many of us really know the constellation of Camelopardis, Vulpecula or Pyxis? Can you recognize all the constellations that are visible from your latitude? The trouble is with the smaller and fainter constellations is that they don't stand out anywhere near as much as their more famous cousins, as the stars they contain are much fainter. The faint pattern of stars is not as instantly recognizable so doesn't linger very long in the memory and unfortunately are easily forgotten. It is always well worth your while learning some of the more indistinct constellations, especially if you are hunting down some faint object which might be nestled amongst its stars. And keep going back to reinforce the memory.

The constellations are quite often depicted on maps and planetarium software with lines connecting the brighter stars, so that the patterns can be more readily recognized. There are a different number of ways in which these lines can be drawn, the way in which these stick figures are drawn differently often result in completely different shapes being drawn depending on the source being referenced. These range from the more traditional outlines, modern outlines or others. Many of the planetarium programs available for the computer enable the user to change these settings for their preferred method. There really is no right or wrong way of

D. Eagle, *From Casual Stargazer to Amateur Astronomer: How to Advance to the Next Level*, The Patrick Moore Practical Astronomy Series, DOI 10.1007/978-1-4614-8766-1_5, © Springer Science+Business Media New York 2014

joining these stars up as long as the observer uses a consistent source to aid them in recognizing them.

Though this is purely personal choice the author prefers to use the many of the depictions shown by H.A. Ray in his classic book, The Stars: A New Way To See Them. Houghton Mifflin, 2008. Originally published in 1952 it is still in print.

5.1   Star-Hopping

When you are out observing, the best way to remember your way around and identify the star patterns is largely due to repetition. Keep going back and re-finding your familiar star patterns. Use these to re-locate your favorite constellations and deep-sky objects. Use these to guide you to some of the less familiar patterns of stars. Once you have done this a number of times, the faint constellations and objects that eluded your gaze for so long will soon also become familiar friends. As a result you will soon be able to find your way around any part of the sky whatever the time of year and be the envy of your fellow observers.

Once you have gained a solid foundation knowledge of the layout of the sky, that familiarity acts as a useful starting reference point to find harder and fainter objects. Learning how to star-hop effectively, you will soon be finding your way through the maze of faint stars, picking up objects "well off the beaten track" with seemingly very little effort at all.

GOTO or not GOTO

In recent years computer controlled telescopes have become much more affordable to the amateur astronomer. These are designed to GOTO the object of interest using a handset of some type. There is a lot of debate amongst amateur whether or not a GOTO system should be used or not. It is the author's opinion that they are extremely useful in finding faint objects and centering them in the field of view, especially useful when imaging. But one thing to note that in many cases, especially in small telescopes, the database contains far too many objects, many of which will be too faint to be visible through the telescope even under the most ideal conditions. In some cases the expense of the scope is in the electronics, with a compromise on the quality of the optics. With observing, the quality of the optics is everything. Electronics is also something else that will probably go wrong in the dark when trying to observe. Use GOTO with caution and don't be afraid to learn your way around the sky yourself. You really will not regret doing so.

5.2   Learn to Star-Hop

Star-Hopping is a technique where we start off with fairly bright stars and use patterns in the stars to guide us to our intended destination. Let us start with a very easy example to practice at finding a very bright object using the star-hopping technique.

Exercise 5A: Star Hopping to M31

In the fall sky in the northern hemisphere, the constellations of Perseus and Andromeda are very proudly on display, located on the meridian around midnight, and visible from most of the world. The Great Square of Pegasus is a distinct asterism of four stars. Using just your naked eye, find this square of stars. Incidentally, the top left star (all instructions are now as seen from the northern hemisphere) Sirrah, is now actually Alpha Andromedae.

From this square of stars take the top edge of the square and carry the line on towards the left (east) and up. Just under half the distance of the top of the square, you should come to another reasonably bright star. Slightly left and up again, taking a slightly longer journey, you will come to another star of similar brightness. This is Mirach or Beta Andromedae. When you have found this star, turn 90° to the right. You will then see two fairly bright stars leading away. Aim for the second star and gaze at this star. Just above and to the right of the second star there is a faint smudge, which marks the location of M31, The Andromeda Galaxy. In five hops we have found our quarry.

OK, so this is a fairly easy object to find, and very bright, easily visible in binoculars, if not the naked eye, but it does serve to show just how easy star-hopping can be. Finding much fainter objects is really just as easy, but observing them once found is likely to be much more of a challenge. Below we will look at how to find fainter objects using this technique.

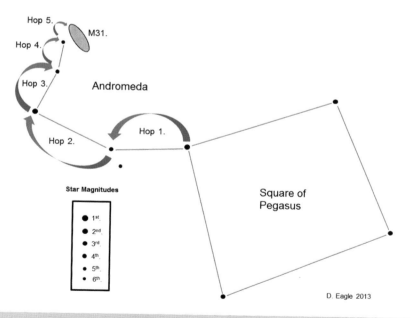

Fig. 5.1 Star hopping to M31 from The Square of Pegasus (Courtesy of the author)

Star-Hopping in Practice

As we have seen star-hopping can get us to our destination really easily if the stars are bright enough and the object is easy to see. But it is just as easy using fainter stars and leading the observer to fainter objects. It is very much like finding a route on a driving map. Let's say you want to get to a tiny little village in the middle of the countryside. To find your best route you will choose the major highways first and plan the most direct journey from there. The less complicated the route you choose, the easier your drive will be. As you get closer to your destination, you will come off the major routes and start going down much smaller roads, eventually arriving at your destination. You will be constantly on the look-out for conspicuous landmarks along the way, perhaps a distinct building or a statue. If you frequently travel the same route time and time again, you will soon know your way without really thinking about how to get there and start to know much of the surrounding area as well.

Finding objects in the sky is very much the same. There are maps to show you where the object is located in the sky, either printed or on a computer screen. There will usually be a fairly bright star close to your target, so start your journey there. Is that bright nearby star in a familiar constellation? So starting out on the first steps of your journey should be easy if you start here, using it as your first landmark. Unless, of course, the object lies in one of the much fainter and less familiar constellations. Here your accrued knowledge of the sky will hold you in good stead as you will already have a head start if you know these lesser patterns of stars. Once you have found your constellation and star of choice, center that bright star in your field of view. We have already found our major highway, now we need to head off in the correct direction towards our quarry. Which direction do we need to move away from this star to home in on your object? North, South, East or West? Make sure that your map is orientated to the same as your eyepiece field of view. If your telescope reverses the field, then the orientation will be different and possibly confusing. This is more confusing if you are using a star diagonal on a refractor. (See Fig. 4.5).

Does your detailed map (you did print one off and bring one out, didn't you?), show you any distinctive star patterns which will help you orientate yourself in the right direction? The eye is very good at picking up lines of stars, patterns and shapes, so find something on your map that is in your field of view that you can identify to lead you in the right direction. Is there a distinct triangle of stars visible close to your target? Does a point of this triangle point towards or away from your target? Are there two stars close together near your object? Are there any brighter stars visible around your object.

All these clues will help to lead you on in the right direction. Already we have turned off the major roads and are rapidly heading down side roads towards our target. If you get lost, re-trace your steps, go back to where you knew you weren't lost and had an easily recognized pattern of stars. It should then just be a case of working your way back through your selected star patterns. Don't despair, you will get there.

Keep repeating the exercise, slowly edging your way forward until the object you are seeking creeps into your field of view. Once you've got there, can you see

your object? If not, are there any distinctive star patterns shown or
you can identify which will prove you are looking at the correct ?
correctly identify the field stars and still cannot see your object, c
why? Is it too faint to be seen through your scope. Did you chec..
telescope collected enough light to see it? Is the Moon or light pollution interferi...
with your observation?

If you did find your object you wanted to observe, rest there a while and enjoy
it. Take it all in. You've taken time to get to this point and you've enjoyed the journey. Now you're here really try and make the most of the view.

Observing takes time. Now you've arrived at you target object, don't just have a
quick view and move on. Scrutinize the object carefully. Take your time to get to
know the object. Take in as much as you can. Is there any sort of shape to it? Does
it have different parts that are of different brightness? If observing a galaxy are
there any new stars visible within it?

So you made it and found your object. Was it worth the journey? Did you enjoy
your view of the object you sought? Would you visit again? Only you can answer
those questions. Don't forget to make some notes as you go, so you have your permanent record.

5.3 Positions of Objects on the Sky

Objects are set in positions in the sky and there are a number of different coordinate systems that the observer will come across and will need to know.

Knowing an objects co-ordinates on the sky enable observers to place a celestial
object in a precise location. So if you want to go back there again, direct another
person to the same spot of a favorite celestial object, or just found out about a new
discovery, you just need to know its current co-ordinates. Stars and deep sky objects
only change in position extremely slowly, so their positions stay fairly static. Solar
system objects move relatively quickly as they orbit the Sun, so their positions are
continually changing.

Celestial Coordinates (Equatorial Coordinates)

The most familiar is the celestial coordinates. This is the equivalent of the role that
Latitude and Longitude play on the Earth's surface. This is the Right Ascension and
Declination of an object on the celestial sphere.

Right Ascension (RA)

This is measured from a point called The First Point of Aries. This is the point on
the celestial sphere where the Sun's apparent yearly path across the sky (called the
Ecliptic) crosses the celestial equator at the Equinox in March. This marks $0°$ in
Right Ascension. Moving Eastwards from this zero point, the sky is divided up into
24 h (h) segments. Each Hour is sub-divided into 60 min of arc (m), and every
minute of arc is further sub-divided into seconds of arc (s).

clination (DEC)

This is the position of the object north or south of the Celestial Equator. This is measured in degrees. Each degree is divided into 60 min of Arc ('), each minute of arc further divided into 60 s of arc ("). North of the celestial equator is given a positive, south of the equator negative.

Objects lying on the celestial equator are at a declination of 0°. The north celestial pole is at a declination of +90°, and the south celestial pole lies at a declination of −90°. Under this system an object on the sky, such as Sirius can have its position written as: RA 06 h 45 m 08.9173 s Dec. −16° 42′ 58.017″. This enables the observer to specify exactly where to look for an object.

Unfortunately, due to the effects of Precession (and to a lesser degree Nutation), results in the position where the first point of Aries slowly moves over time. The progression is westwards across the celestial sphere. The so-called First Point of Aries has now shifted into the constellation of Pisces. In about the year 2600 this will cross into the constellation of Aquarius. Hence the origin of the popular song lyrics "The Dawning Of The Age Of Aquarius". What this does mean is that position of coordinates on the celestial sphere are slowly changing over time.

Fortunately modern computers can cope with this constant predictable change, but in practice it just isn't practical to keep changing the coordinates. Many maps and published catalogues choose a particular epoch for their data. So some of the older atlases and catalogues were written to use the epoch of 1950. This was consistent for much of the twentieth century. As the twentieth century drew to a close the epoch of 2000 was later adopted. Most atlases are now drawn and positional data is usually published using this epoch. This isn't something that will cause the amateur too many problems, but it is something to be aware of. It is always best to make sure that your star atlases and references are as up to date as possible.

Electronic devices such as telescope drives and planetarium software often use the current date as their epoch as they are able to calculate these small changes in position over time very easily.

Horizontal Coordinates

This is the coordinate system that is used to describe the direction and height you are facing according the observers horizon. This is using Altitude (alt) and Azimuth (az) which should be familiar to you as alt-az. This measurement will be different for every observer, particularly so if they are separated by large distances.

Altitude

Altitude is measured in degrees (°) from the horizon. An object directly on the horizon will have a zero value. An object directly above the observers head (at the Zenith) will have a value of +90°. The degrees are also divided into 60 min of arc (') and every minute divided into 60 s of arc ("), so the positions of objects can be measured extremely accurately. Something, which has not yet risen above the horizon (if indeed it does), will have a negative value. The larger the negative value, the further the object is below the horizon. The point directly beneath the observers feet (−90°) is known as The Nadir.

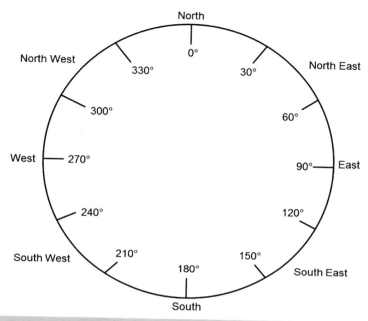

Fig. 5.2 The azimuth scale (Courtesy of the author)

Azimuth

Azimuth is measured according to the direction in which the observer is looking. This is measured from the north point on the horizon which has a value of zero degrees (0°), turning clockwise. Therefore the Eastern horizon has a value of 90°, south 180° and west 270°. This value can also be further subdivided into decimal values so that greater accuracy can be obtained. There are therefore 360×60×60 arc sec diving the whole sky. This adds up to 1,296,000 s of arc for a 360° radius.

The alt-az values of the positions of satellite flares are often used on satellite prediction Websites such as Heavens Above.

www.heavens-above.com

Galactic Coordinates

These are used to describe the position of objects within our visible Milky Way galaxy. The galactic coordinates are measured using a measure of galactic latitude and longitude from the plane of the Milky Way. This is inclined by 63° to the celestial equator. The zero longitude point lies towards the galactic center in Sagittarius, with degrees of longitude measured eastwards across the sky from this point. Latitudes north of the galactic plane have a positive value, south being negative. The north galactic pole lies within the constellation of Coma Berenices. The south galactic pole lies within Sculptor.

Ecliptic Coordinates

These measurements are not used very often. As most of the major planets lie fairly close to the ecliptic, this measurement is used to indicate the position of planets along the ecliptic. The unit of measurement used is degrees. This is measured from the zero point at the First Point of Aries, where the ecliptic crosses the equator in Pisces northwards. As the planets do not always lie exactly on the ecliptic a measurement equivalent to latitude is also used to describe how far north or south of the ecliptic the planet lies.

Another couple of coordinate systems that the observer might encounter and are the Heliocentric and Barycentric coordinates. These are used to give the position of objects as observed from the center of the Sun or in the solar system. The observer will rarely need to know about these coordinate systems or how they are used unless they specialize.

5.4 Asterisms

Asterisms are small patterns or collections of stars that aren't really true constellations. They can reside within a constellation, the Big Dipper (or the Plough) being the most famous example. These seven main stars are part of the much larger constellation of Ursa Major, the Great Bear.

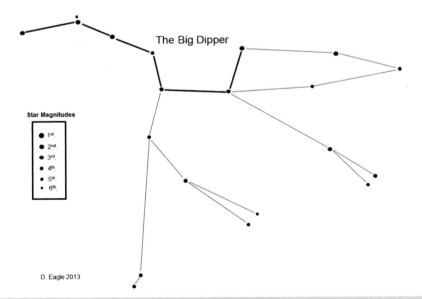

Fig. 5.3 The big dipper in Ursa Major (Courtesy of the author)

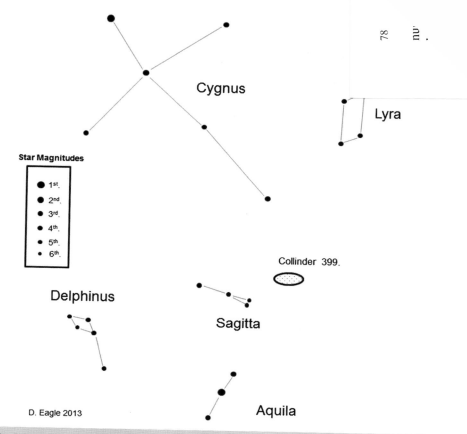

Star Magnitudes

- 1st
- 2nd
- 3rd
- 4th
- 5th
- 6th

Cygnus

Lyra

Delphinus

Collinder 399

Sagitta

Aquila

D. Eagle 2013

Fig. 5.4 Finder chart for the Coathanger, (Collinder 399) in Vulpecula (Courtesy of the author)

They can also be part of an extended pattern of stars, like the Summer Triangle, comprised of the first magnitude stars Deneb, Altair and Vega. The indistinct constellation of Vulpecula contains one of the most splendid asterisms. Collinder 399, or Brocchi's Cluster is also well known as The Coat Hanger. One glance at this splendid object in a pair of binoculars will immediately show you why it has this name.

Although they don't have any scientific purpose and are usually just a random scattering of stars, it can be quite a lot of fun just to look some of them up for your own amusement, especially on nights where the Moon would otherwise wash out most other objects.

There are quite a number of these objects scattered around the sky. They range from a huge pattern like the Big Dipper or The Summer Triangle, marked by Altair, Deneb and Vega, to really small objects like the open star cluster NGC2169 in the northern part of Orion. This really does look like the stars are tracing out the

...ber 37. The much larger Kemble's cascade in Camelopardus is also worth hunting out, and using a wide angle field of view or binoculars look out for the Smiley Face close to M38 in Auriga. When you do spot this make sure that you smile back at him (or her). In your meanderings round the sky you will find many more patterns of stars that will entertain you and guide you on in your observing adventures.

For further information on Asterisms:

David Ratledge's guide to Asterisms.
http://www.deep-sky.co.uk/asterisms.htm
Saguaro Astronomy Club's Guide to Asterisms.
http://www.nightskyatlas.com/asterisms.jsp

5.5 The Brightness of Stars

When we casually look up at the sky we soon start to see that not all stars are the same. Sure, they vary in brightness, but this is not always a property of the star itself. This is often confused by their different distances (see below) but we soon realize that many of the brightest stars also show different colors. Compare the cool red hue of the red giant star Betelgeuse in Orion's shoulder with the hot blue color of Rigel at his foot. This color difference is due to the surface temperature on the star itself. Relatively cool stars are red, hotter stars appearing white or blue, with a wide range of colors between. Some stars are variable and appear to change in brightness over a period of time. Other stars appear to be associated with one another forming double stars. There are even multiple star systems, such as Epsilon Lyrae.

As an observer you can really delve deeper into observing the stars themselves and learn a little bit more about their properties. Humans always like to catalogue and classify observed objects and the brightness of stars is no exception. To measure the brightness of a star the magnitude scale is used. There are however a number of different magnitude scales in existence frequently encountered by the amateur observer:

Apparent Magnitude (m)

This is the brightness of an object as measured by the observer. This is only a measure of the apparent brightness of a star as seen from our vantage point on Earth. It does not tell us the real luminosity of the objects themselves. So a very brilliant star at a great distance can look quite dim. A much dimmer star, which is located, relatively much close to the Earth could look really bright.

The Magnitude Scale

The Greek Astronomer Hipparchus devised a system over 2,000 years ago called the magnitude scale. He classified the brightest stars as being of first magnitude. The faintest stars he could see, considering his skies were much darker than ours, were around sixth magnitude. So the faintest stars were assigned a higher number. This sounds a little topsy-turvy until you do what Robert Burnham suggested in his celebrated Celestial Handbook and substitute the word magnitude for the word class. When you look at it in this way you start to refer to them as first class stars, second class stars and so on as the stars get fainter. This brings the magnitude scale a little bit more into perspective.

The magnitude scale is logarithmic, meaning each whole step in magnitude is 2.512 times brighter than the next. A 1st magnitude star is about 100 times brighter than a 6th magnitude star. This is a difference of five magnitudes. This is calculated as shown below.

$$2.512 \times 2.512 \times 2.512 \times 2.512 \times 2.512 = 100.02, \text{ or as it is normally written:}$$
$$2.512^5 = 100.02$$

A 3rd magnitude star is 6.3 times fainter than a 1st magnitude star, a difference of two magnitudes.

$$2.512 \times 2.512 = 6.3 \text{ or } 2.512^2 = 6.3$$

In some cases there are objects that are even brighter than first magnitude. There are stars that are classified as being of zero magnitude (wrongly suggesting it has no brightness). Stars brighter than zero magnitude are given a magnitude value that is negative. Here are the magnitudes of some common celestial objects for comparison (Table 5.1):

Table 5.1 Average magnitudes of some celestial objects	
Object	Magnitude
Sun	−27
Full moon	−12
Venus at brightest	−4.4
Arcturus	−0.04
Vega	0.03
Polaris	1.99
Pluto	13.9

Using telescopes the amateur astronomer can now detect objects way below the sixth magnitude naked eye limit of objects on Hipparchus' original scale. The bigger the aperture of the telescope, the fainter the objects that can be seen. The Hubble Space Telescope can detect stars down to 30th magnitude. This is a measure of the apparent brightness of the stars as viewed from the Earth.

Absolute Magnitude (M)

This is the brightness a celestial object would have if it was placed at a distance of 10 pc (32.6 Light Years). This enables the observer to compare the brightness of one star to another. A star with a high absolute magnitude value would be much more powerful than one lower down the scale.

Visual Magnitude (Vo)

This is a measurement of a celestial body as seen by the human eye, which has a peak in sensitivity at a wavelength of around 5,600 Å.

Photographic Magnitude

This is a measurement of a stars brightness at the peak sensitivity of traditional film photographic emulsions. This is down at the blue end of the spectrum. Largely outdated now, but still appears in the literature from time to time.

Apparent Photographic magnitude = mpg.
Absolute Photographic magnitude = Mpg.
Today most magnitude measurements are performed electronically using sophisticated CCD cameras.

5.6 The Bayer Designation

Johann Bayer published his Uranometria star map in 1603. He assigned Greek letters to individual stars within each constellation. Many people today assume that these letters were assigned according to the brightness of the stars. In many constellations this is true, but at the time magnitudes of stars could not be very accurately measured. His lettering assignment was quite often done in a less regimented manner. It could be done by the position of the stars within the constellation, or using the stars positions in right ascension. Today we still use these to identify the brightest stars and many people still cling to the belief that alpha designates the brightest star, beta the second brightest and so on. Let's go on to disprove this myth using the familiar asterism of the Big Dipper.

Exercise 5B: Compare the Brightness of the Stars in the Big Dipper

Here are the stars in The Big Dipper listed according to their Bayer letters (Table 5.2):

Table 5.2 Names of the seven brightest Big Dipper stars listed according to their Bayer designation

Star name	Bayer letter
Dubhe	Alpha (α)
Merak	Beta (β)
Phad	Gamma (γ)
Megrez	Delta (δ)
Alioth	Epsilon (ε)
Mizar	Zeta (ζ)
Alkaid	Eta (η)

You would assume from this table (had Bayer assigned the Greek letter according to brightness) that Dubhe would be the brightest star, Merak the second brightest and so on straight down the table. Is that indeed the case? Next time it's a clear night go and have a look for yourself.

Print out the map of The Big Dipper (Fig. 5.3 – Download a copy from here: www.eagleseye.me.uk/Guides/Dipper.pdf). Print it out and take it outside, compare the brightness of all seven stars.

Try and estimate their relative brightness. Assign a Greek letter to each of the stars in order of brightness and mark them on your chart by each star. Label what you estimate to be the brightest star as Alpha, the second brightest Beta and so on right the way through all seven stars. How does your new Bayer designation compare to how they are labeled in star maps?

You will have noticed that Megrez, located where the handle meets the bowl, is definitely the faintest star of the seven. If the Bayer letter is to be believed, then it would be the 4th brightest star.

So we have now disproved that the Bayer letter always indicates the star brightness in a constellation. Bayer actually assigned his Greek letters in Ursa Major according to their right ascension, not brightness.

The exercise you have just done is also useful in looking the accuracy of your perception of brightness. Look up the measured magnitudes of the Big Dippers Stars in Table 5.3 below. Did you get all seven stars listed in their correct order of brightness?

Table 5.3 A list of the magnitudes of the seven stars in The Big Dipper

Star name	Magnitude
Dubhe	1.81
Merak	2.34
Phad	2.41
Megrez	3.32
Alioth	1.76
Mizar	2.23
Alkaid	1.85

It might also be useful to repeat this same exercise on a night when conditions are different. e.g. a slightly misty night, or a night when the Moon is near to full to see if you get different results. Or how about putting the exercise to a group of people to try the exercise individually and collating those results to see how accurately these estimations become when more data is collected in this way?

5.7   Limiting Naked Eye Magnitude

How faint a star can you see with the naked eye from your observing site?

This is used as a measure of how dark your skies are and how your skies may be affected by light pollution or general seeing conditions.

It is often quite useful to be able to judge just how dark are your skies and the effect (or not) that light pollution or faint haze has on the amount of stars you can see on any given night. The darkness and clarity of the skies can be different each night and can even change minute by minute. A good dark, clear night will allow you to see much fainter stars with your naked eye, and this is a good judge to measure the darkness and seeing of your sky.

So let us see how you can estimate how faint a star you can see with your naked eye from your location. This is usually known as the limiting magnitude. There are a couple of areas of the sky where the magnitude of the surrounding stars are fairly well known. By looking at that area of sky you can identify the stars that are visible using just the naked eye.

In the northern hemisphere we have what is known as the North Polar Sequence. This was devised by Alistair McBeath in 1988. There are a set of maps with a selection of stars printed on the map with their respective brightness marked. By identifying which stars are visible from your location, you can get a quick measure of how dark your skies are.

Download and print the charts from here: http://www.jjdash.demon.co.uk/north-polar-sequence.html

Fortunately this area of sky, being circumpolar, is visible all the time in the northern hemisphere. Unfortunately in the southern hemisphere there isn't quite as good an equivalent sequence of stars around the southern celestial pole. Another area of sky that can be used to determine your limiting naked eye magnitude is within the Square of Pegasus. This area of sky is visible from most of the inhabited world. The square itself is very easily identified. To measure the darkness of your skies count the number of stars that are visible to the naked eye within the square. If no stars are visible then we can definitely say that the sky background is very bright and only stars above magnitude four are visible. If you can count nine stars, the faintest naked eye star is of magnitude 5.75. If you can count more than nine stars within the square then you are fortunate in having extremely dark skies.

All details are given on the following Web page with downloadable pdf available of the sequence: http://freestarcharts.com/index.php/stars/17-guides/stars/17

5.8   Estimating Star Magnitudes

Let us now move things on a bit more and start looking at the brightness of stars themselves. The easiest and best way of doing this is to estimate the brightness of a star by comparing it to others. We have already had a quick comparison of the relative brightness of the seven stars in The Big Dipper, but let's refine things and make it a bit more accurate and estimate a stars magnitude.

Exercise 5C: How to Estimate the Magnitude of a Star

Let's start by estimating the magnitude of γ Ursae Minoris. Look towards the North and find the constellation of The Little Bear, (Ursa Minor). It is conveniently on display at all times of the year in the northern hemisphere. This constellation provides a good range of star magnitudes for excellent practice in estimating star

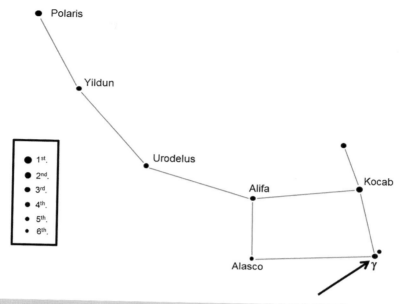

Fig. 5.5 Map of Ursa Minor showing location of gamma (Courtesy of the author)

brightness using the naked eye. The seven main stars of the constellation are marked in the map. We will try and estimate the magnitude of γ Ursae Minoris (arrowed) using just your naked eye.

Look carefully at the stars surrounding γ. You need to identify two that are fairly close to the star you are trying to estimate. These will be our comparison stars. In this case, two of the stars, Kocab (β)and Alifa (ζ) conveniently lie either side of γ in magnitude.

Observe the three stars carefully and estimate how far from each star γ varies. Is it just fainter than the brightest comparison star, just brighter than the fainter star, or somewhere in between? Try and determine where that star lies in brightness on a scale of 1–10 between the two stars.

Have you got a measurement yourself? In this instance you should be able to see that the brightness of this star is almost exactly midway between our two comparison stars, so we assign a score of five on our scale,

The magnitudes of the two comparison stars we have just compared it to are given as 2.1 and 4.3. If our estimated star is midway between these two stars it gives us an estimated magnitude for γ of around 3.0. (γ's magnitude is in fact 3.05 which is extremely close to our approximation).

Carefully observing stars in this way will enable you to calculate magnitudes of other objects seen in the sky. This skill will also come in very useful when estimating the brightness meteors, minor planets and comets (more of which later). This method is most certainly the best method used for our next subject, Variable Stars.

5.9   The Colors of Stars

Let us start now to look at another property of stars. Their colors.

As well as looking pretty, the color of each star tells you a lot about the processes going on in the star and the type of star it is. The color is usually representative of the temperature of the stars outer "surface". The cooler the star, the redder it looks. The hotter the star, the bluer or whiter it will look. The color will tell us something about its history, or even where it might be along the evolution of its life.

5.10   The Hertzprung-Russell Diagram

In 1910, two astronomers, Ejnar Hertzprung and Henry Norris Russell, plotted stars on a graph with luminosity on the vertical axes, and spectral types along the horizontal axes.

When plotted on this diagram most stars appear to show a relationship trending along a line from bottom right, to upper left. This has been interpreted to show that stars spend the majority of their lives located somewhere along this line. So the line represents a generic evolutionary track. This has become known as The Main Sequence.

More about the Hertzprung-Russell Diagram found at these resources here:

http://zebu.uoregon.edu/~soper/Stars/hrdiagram.html
http://outreach.atnf.csiro.au/education/senior/astrophysics/stellarevolution_hrintro.html

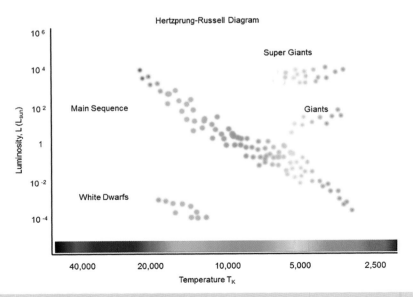

Fig. 5.6 Simplified version of the Hertzprung-Russell diagram (Courtesy of the author)

5.11 Investigating the Colors of Stars

Point your telescope at the star in the head of Cygnus the Swan. This star Albeiro, also known as ß Cygni, is a beautiful double star system. No doubt this was one of the first double stars you visited when your interest in astronomy was first kindled.

What makes this double star so special is that as well as being very bright and easy to find, the two stars are very different in color. Seeing such contrasting colored stars in such close proximity really brings out the colors, especially if they are so different. The brightest star is yellow or orange, the fainter being bluish, (or something like that). Of course the color you see will depend on your personal perception.

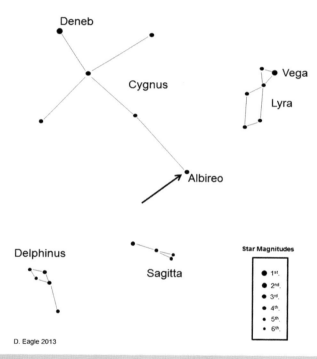

D. Eagle 2013

Fig. 5.7 Finder chart for the double star Albireo in Cygnus (Courtesy of the author)

What colors do you observe when you look at these stars? Do the colors appear different if viewed through a different aperture telescope? Does haze in the atmosphere affect the colors you see?

When the time of year is right, take a look at Betelgeuse in the shoulder of Orion. Take in its luscious ruddy hue, before heading south to look at Rigel in Orion's foot. The difference between the two contrasting types of stars is instantly obvious (even to the naked eye observer).

re observing some of your deep sky objects and open clusters, take
ook at the stars in the field of view. See if you can pick out stars of
's you may have otherwise overlooked.

ample of this is the double cluster in Perseus. Also known as "The
ιe" it is a rich field of stars with the two individual clusters virtually
running . . one another. Between the two clusters (being old clusters of stars)
there are a number of old red stars nestled between them. See if you can pick some
of them out while observing the scattering of jewels.

Exercise 5D: Picking Out Red Stars in the Sword Handle

The Sword Handle in Perseus is two open clusters lying side-by-side. Situated
between the constellations of Perseus and Cassiopeia it is a beautiful field of stars
in a small telescope, like diamonds spread onto a velvet cloth. It is visible to the
naked eye as a small smudgy patch, and shows extremely well in binoculars. But
let's investigate these clusters a little better. Use whatever telescope you have to
look very carefully at the stars scattered around the field of view. Can you actually
make out where one cluster ends and the other one begins? Keep scanning around
and pay careful attention to the individual stars themselves. Did you notice that
there are a number of red stars scattered amongst the other stars?

If you didn't see these initially, look again and they will certainly pop out at you
once you know this fact and you do start being more aware that they exist and take
the time to take it all in.

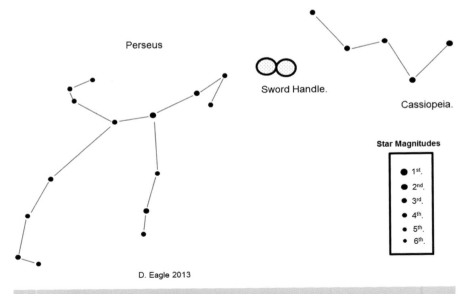

D. Eagle 2013

Fig. 5.8 Finder chart for the sword handle in Perseus (Courtesy of the author)

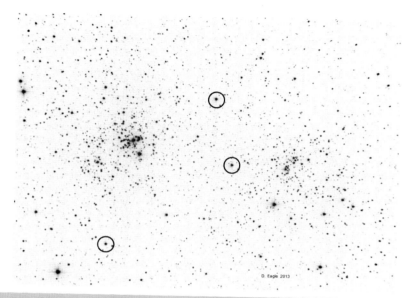

Fig. 5.9 Negative image of the sword handle in Perseus with three red stars marked. Have a look, can you find any more? (Courtesy of the author)

5.12 Simple Spectroscopy

To develop your analysis of starlight further the next step is to arm yourself with a relatively cheap diffraction grating like a Star Analyser to produce a spectrum of stars and other objects. It simply screws into the barrel of the eyepiece and produces a spectrum of any object that shines through it. It will need some further investment, so am not going to go into more details in this book.

Star Analyser

This relatively cheap diffraction grating mounted in a standard inch and quarter filter cell can be used successfully to produce spectra of stars and other celestial objects.

http://www.patonhawksley.co.uk/staranalyser.html

Astronomical Spectroscopy for Amateurs. K.M. Harrison. Springer.

Spectroscopy: The Key to the Stars: Reading the Lines in Stellar Spectra. K. Robinson. Springer. 2007

Grating Spectroscopes and How to Use Them. Ken M. Harrison. Springer.

Practical Amateur Spectroscopy. Stephen F. Tonkin. Springer.

5.13   Star Catalogs

As well as the Bayer designations shown above, stars have also been catalogued a number of times over the years. Some are specialized catalogues such as the PPM and Hipparcos catalogues list the proper motions of stars.

Many of the brightest, or more interesting stars have been given their own unique names. Many of the brighter stars, like Sirius and Arcturus, have been passed down to us from antiquity. Others have been named more recently as their interesting characteristics have been identified by astronomers, e.g. Barnard's Star. Most observers always use a stars proper name (if it has one) so it is always useful to learn most of the named stars and where they can be found in the sky.

Over the years quite a number of star catalogues have been developed. Many of them, are specialized listings of particular types of stars, Doubles, Variables etc.

The ones that the amateur most frequently encounters are the Smithsonian Astrophysical Star Catalog (SAO) and the Guide Star Catalog (GSC). The GSC catalogue is frequently used by planetarium software to plot stars down to magnitude 15. The latest version has star positions down to magnitude 21 and is used to help guide the Hubble Space Telescope, hence its name.

Chapter 6

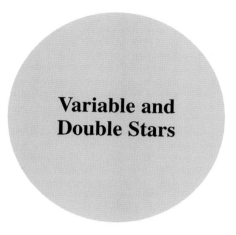

Variable and Double Stars

6.1 Variable Stars

The bright variable star Beta Persei, or Algol, the Demon Star has been known to vary in brightness for many thousands of years. There are many over variable stars scattered around the sky. Their variability can be due to a variety of different causes. Algol itself is an unresolved double star system. When the fainter companion moves across the face of the brightest star, the amount of light reaching the observer is reduced. A much smaller reduction on brightness is seen when the fainter companion passes behind the brighter star. Another variable star Epsilon Aurigae, located close to Capella, although itself a double star system, its variation in light output is mainly due to a cloud of dark material moving around the two stars. As this dark material orbits the stars it periodically stops some of the starlight from reaching us on Earth. This change is reasonably predictable, but changes do take place over time as the material moves around. So like many things that change in the sky, it is always worth keeping track of what is happening.

6.2 Naming of Variable Stars

The naming system for variable stars was developed by F. Argelander in the 1800s. Stars that are discovered to be variable which do not have designated names such a proper name or a Bayer number or Greek letter are designated letters. This starts at the letter R right through to Z. So a new variable in a particular constellation will be called, say R Doradus. Once the letters have been used a double letter is used.

D. Eagle, *From Casual Stargazer to Amateur Astronomer: How to Advance to the Next Level*, The Patrick Moore Practical Astronomy Series, DOI 10.1007/978-1-4614-8766-1_6, © Springer Science+Business Media New York 2014

So we get variable star names like RR Cygni. Once ZZ is reached the alphabet from A to Q is then used. The letter J is not used. When the system was first devised the number of variable stars known was very small, so was thought at the time to be able to cope with the number of variable stars likely to be identified. Unfortunately as our techniques became much more sophisticated we can identify very small changes in variability and in very faint stars. This has produced a huge catalogue of variable stars. Once this system of 334 possible combinations is used up a V designation is used with a number. E.g. V5558 Sagittarii.

Each variable star has its own personality, varying much differently to any other variable star. The brightness variation can range widely, from stars whose variability is scarcely noticeable, to stars where their variability is outrageously flamboyant, ranging across many magnitudes.

6.3   Types of Variable Stars

There are seven main groups of variable stars known. Variability is divided into two main types depending on its cause, either intrinsic and extrinsic variability.

Intrinsic Variables

Intrinsic variability means that the reason for the variability lies solely with the varying light output of the star itself.

These are stars which vary their output of energy, so the variability in brightness is directly associated with the activity of the star itself. These include Mira, Cepheid variables and Semi-Regular variables, which are divided up into a number of sub-groups.

Eruptive Variables

These stars vary due to violent activity like flares occurring regularly on their chromosphere or coronae. This interacts with the surrounding interstellar material producing bursts of light that can be seen from Earth.

Pulsating Variables

The stars in this category are expanding and shrinking periodically which increases or decreases the surface temperature and output of energy from the star.

These stars usually show a fairly predictable cycle to their variability making them very easy to predict future behavior. This group includes the Mira and Cepheid variable stars. Cepheid's are extremely important stars as their period of variability is directly linked to their brightness. They are therefore used as "standard candles". If the star is identified as a Cepheid variable and the period of variability of the star can be determined, it tells the observer its true brilliance. By working out the difference in apparent magnitude that the star appears in our sky, its distance can be calculated.

Rotating Stars

These stars vary not due to changing output of energy from the star but produced by a non-uniform surface brightness. This could be due to hotter bright patches or dimmer patches that are heavily sun-spotted. As the stars revolve, brighter or dimmer areas are bought into view. This type of variability is also caused by close orbiting binary stars. As they orbit around one another so closely, their disks are non-spherical. Their apparent constantly changing shape as they orbit results in the variability in light reaching the observer.

Cataclysmic Variables

These are variables that display dramatic changes in brightness. These types of variable stars used to be known as novae (New). It is believed that they are composed of two stars, one a white dwarf primary star and a mass transferring secondary. Orbiting so close to the primary, matter is transferred causing outbursts of UV and X-ray radiation. The accretion disk is very unstable and outburst are regularly recorded. It is thought that systems like this eventually produce a Type Ia supernova.

Flare Stars

Variable stars that show a dramatic increase in brightness. This is thought to be due to strong solar flares on the photosphere of the star causing a sudden flash of light. These only last for a matter of a few minutes.

Extrinsic Variables

Extrinsic variability means that star itself is constant in energy and light output, but something about its environment causes the amount of light reaching the Earth to vary. A good example of this type of variable star is Epsilon Aurigae. This star has a period of over 27 years and variability is caused by a dust cloud surrounding the star. Movements of the dust surrounding the star(s) alternatively reveals and conceals the stars within.

6.4 Supernovae

Not variable stars in themselves, these are the final violent act of dying stars at the end of their lives. Not many supernova have been witnessed in our Milky Way galaxy over the past few hundred years. Being as bright as these events are, many have been witnessed in remote galaxies. In fact the output of light from a supernova can often outshine the whole host galaxy which contains billions of other stars. Supernova are categorized into two main types, type I and type II.

Type I Supernova

There are three main sub-types, mainly identified using spectroscopic analysis:

Type Ia – Displays a strong Ionized silicon absorption line.
Type Ib – No silica, but shows a non-ionized helium absorption line.
Type Ic – No silica line and very little or no Helium absorption lines.

Type II Supernova

These can also be sub-divided according to their spectra. They show strong hydrogen absorption lines in their spectra and display extremely broad absorption lines. This indicates extremely fast expansion velocities.

Type II supernova are classified into four types:

Type II-P – Reach a "plateau" during their light curve.
Type II-L – Displays a "linear" decrease in its light curve.
Type IIn – Displays narrow lines in its spectra.
Type IIb – Spectrum changes as it progresses to become more like Type Ib.

Exercise 6A: Estimating the Distance of a Type Ia Supernova

Type Ia supernovae are extremely useful in measuring the distance of remote galaxies. It is known that type Ia supernovae always reach the same brilliance at their peak. Like Cepheid variables they can be used as "standard candles" to determine distances. Type Ia supernova are extremely brilliant, often outshining their parent galaxy, so unlike Cepheid variables they can be observed across tens of millions of light years. So the distances to very distant galaxies can be determined using them.

So let's work out how to determine the distance to a distant galaxy if we know the brightness of a type Ia supernova. A typical type Ia supernova has an absolute magnitude of 19. As this is fairly well set in stone, if we observe a supernova in a distant galaxy and measure the actual magnitude that we observe, then the difference in apparent brightness gives us a reasonably good measure of the galaxies distance.

The formula used is:

$$M = 5 + m - 5 \log d.$$

Where M = Absolute magnitude (Magnitude 19.3).
m = Apparent magnitude (Brightness as observed from Earth).
d = Distance in Parsecs (1 pc = 3.26 light years).

A Type Ia supernova (SN 2011fe) was seen in M101 in 2011. The brightest apparent magnitude it attained was +10. This was easily viewed by amateur astronomers for many weeks, even bright enough to be seen in binoculars. We now have all the information we need to rewrite the equation to start to calculate the distance to M101.

$$19.3 = 5 + 10 - 5 \log d$$

Simplified: $19.3 = 15 - 5 \log d$
The only unknown is now the distance, so the equation can be re-arranged thus:

$$5 \log d = 15 + 19.3$$
$$5 \log d = 34.3$$

This is then re-arranged to give:

$$\log d = \frac{34.3}{5} \quad : \quad \log d = 6.86$$

The anti-log of 6.86 from a scientific calculator gives us 7,244,359 pc. As there are 3.26 light years in a single parsec, our final calculation gives us:

$$7,244,359 \times 3.26 = 23,616,612$$

So our final calculation gives us a distance to M101 of 23.6 million light years. The currently accepted distance of M101, determined by the latest Hubble Cepheid variable measurements is around 22 million light years. So our rough calculation is really not too far out. So it's not bad from a fairly casual observation made at the telescope.

The real trick in using type Ia supernova to determine their distance is to catch them long before they reach maximum brightness, so that we can be sure that we are measuring them at their maximum stage, and we are sure that the supernova is of Type Ia.

6.5   Naming of Supernova

A newly announced supernova is given a temporary number. This reflects the objects Right Ascension and Declination. Thus the supernova that was discovered in NGC 4414 in June 2013 was initially called J12262933+3113383. Once confirmation of the supernova has been established it is then assigned an official number by the International Astronomical Union's Central Bureau for Astronomical Telegrams. This starts with the two capital letters SN, to distinguish it as a supernova, followed by the year of its discovery. This is then followed by a lower case letter, a – z. Once over 26 supernova have been discovered and all the letters are used up, another letter is added. So we can end up with something like SN 2013am. This was a Type II supernova that was seen to appear within the spiral galaxy M65 in Leo. A few hundred supernovae are discovered each year.

Amateur's astronomers do extremely well in discovering supernovae. Tom Boles at the time of writing has so far discovered 154 supernovae in distant galaxies, and counting. It comes at a cost, he monitors as many as 12,000 galaxies a night.

Visit his Web page for more details. http://www.supernova.myzen.co.uk/

The International Astronomical Union's Central Bureau for Astronomical Telegrams list of recent supernovae are listed on their Web page: http://www.cbat.eps.harvard.edu/lists/RecentSupernovae.html

Another useful Web page listing the latest supernovae is maintained by The Rochester Academy of Science: http://www.rochesterastronomy.org/supernova.html

6.6   Variable Star Observing Programs

It is only by observers regularly studying the way that these stars vary in brightness, the range of brightness they change and how regularly these changes occur, can scientists can get a real idea as to what is going on around or in those stars to cause the changes we can see.

Exercise 6B: Estimating a Variable Star

First you need to pick a suitable target. Make sure you pick a star that is easy to observe at the time you make your observations. Have a good finder chart to hand when stalking your prey. Finder charts are available to download for free from the American Association of Variable Star Observers (AAVSO) website: http://www. aavso.org/

If you know the star name a finder chart can be created from the AAVSO Star finer on their Web site. Using a wider field map, Star-Hop your way to the intended target. Make sure you identify the field of view of your variable star correctly.

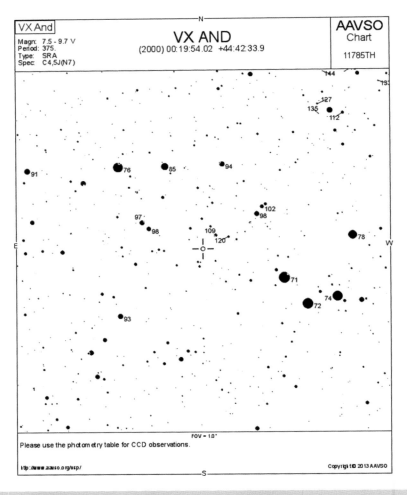

Fig. 6.1 AAVSO comparison star chart (Courtesy of the AAVSO)

When tackling fainter stars in star fields relatively unknown, it can take some time to identify the star and its associated comparison stars. The AAVSO charts usually show field stars that are visible in a telescope around the variable star. Once you have identified field stars included in the variable star chart, switch to using that. The chart is designed to help the observer. A telescope will possibly show some fainter stars than are shown on the chart. This could confuse you initially, but you will soon learn to find your way round, identifying the brightest stars that are shown. The AAVSO charts list magnitudes of comparison stars on their charts that are known not to be variable. The decimal point is omitted on the charts so that it is not confused with a star plotted on the chart, so a star labeled on the map with 45 would be magnitude 4.5, and so on.

Once you have found the correct field of view identify the variable star itself. Once you have found your target, you then need to compare it to some of the comparison stars identified on the chart.

As in the previous exercise, estimating the brightness of a star, the comparison stars you choose should lie either side in brightness of the star of interest at the time you observe it. It is then down to your own judgment to decide how bright the variable star is compared to your chosen comparison stars. Is it halfway between them, or is it somewhere else in between, being closer to one than the other? Make your own judgment but again always be consistent in how you make your comparison. Once you have decided, you now have a comparative estimate of the brightness when compared to these two stars. Knowing the magnitudes of the two comparison stars enables you to obtain a reasonable estimate of the magnitude of our star. In this example we will pretend that there are two comparison stars shown at magnitudes 8.0 and 7.0 magnitude. If our star is one third from the faintest star towards the brightest, this gives us an estimated magnitude of about 7.7. You will become more proficient and more accurate the more times you make estimates. As in all things, practice makes near(ly) perfect.

Of course taking just a single measurement of a star is not really of much scientific use. Only by following the stars over time, as they change in brightness, up and down, can its behavior be properly determined. Pick your variable star of interest with care and you can make its observation into an ongoing project that could last for many years. Follow it over an extended period of time and keep making accurate records of its behavior, you could even confirm possible changes in a known variable stars behavior. In some instances, you might even discover a new variable star.

Once you have obtained a number of magnitude estimates, you can plot those figures into a graph and see if you can see any trend in the fading or brightening of the star in question.

EG Cephei is a near contact binary system, where the two stars orbit almost in contact with one another. Luckily there is an army of willing amateur astronomers ready and willing to participate in estimating the brightness of variable stars and keep an eye out for any unusual behavior. It is quite often amateurs that spot these sort of changes and really do add to the knowledge base to understand the behavior

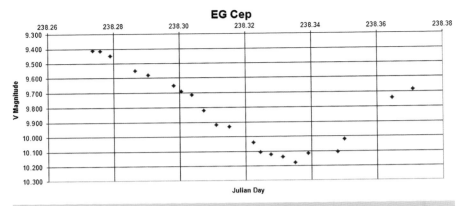

Fig. 6.2 Light curve of EG cephei deduced from amateur observations (Courtesy of Steve Parkinson)

and nature of variable stars. By submitting observations, a vast collection of data can be collated. These combined efforts give a more accurate picture than a single observation or observer ever could. If you are interested in making variable star estimations it is always well worthwhile getting involved and taking part. Your seemingly small contribution really can make a difference to our understanding of the way variable stars develop and change. Remember that even today we are constantly learning something new. Just before this book was submitted for publication, new types of very fast rotating variable stars were discovered in the open star cluster NGC3766.

Do you know the magnitude of the faintest star visible in your telescope? Have you ever tried to find out the faintest star that is visible through your scope? If not, why not? How would you know otherwise if you are hunting down faint objects, if is just too faint for you to realistically see?

As seen previously, the bigger the aperture of the telescope the more light it collects and the fainter stars (and other objects) that can be seen. The lowest magnitude you can see will also be dictated by the amount of stray light pollution, haze in the atmosphere, or the observer's experience.

6.7   Double and Multiple Stars

Double stars are incredibly common in the universe. It is estimated at least half the stars in the sky have companion stars. Double stars are forever changing as the component stars slowly orbit around one another. The speed with which they change is dictated by the distance of the stars from one another and their distance from us. They range from widely separated stars visible in binoculars, to stars that are so close together that their duplicity is only detected by small changes as the stars colors involved in the celestial dance are seen to alternatively red or blue-shift when observed using a spectroscope. We are viewing each star system with their orbits inclined at many different angles to us.

6.8   The Main Types of Double Stars

Optical Binaries

These stars are not associated with one another at all and lie many light years from one another. The only reason they look close together is purely because they lie along the same line-of-sight from our observing location.

Visual Binaries

These are all the double stars that are associated with one another that can be viewed through our optical instruments.

Spectroscopic/Spectrum Binaries

These binary stars are so close together that they can only be detected by using a spectroscope. In Spectroscopic doubles the light from each star is alternatively blue and red-shifted depending on whether the star is approaching or receding from Earth as it orbits the other star. In Spectrum Doubles the resulting spectra doesn't show a shift in the light spectrum but instead reveals a number of dark absorption lines that could not exist in a single star type. So the spectrum must come from two or more different type of star.

Astrometric Binaries

These are stars where the duality of the star cannot be detected other than the motion of the star through space. A slight "wobbling" motion of the apparently single star as it moves across the more remote background stars gives away the presence of another star orbiting around it.

Extra-Solar Planets

Showing just how advanced some of these techniques have now become, it is now possible to detect extra-solar planets around other stars using many of these methods. Something that was deemed to be virtually impossible not too many years ago. This was for many years considered far too difficult achieve. A very tentative paper revealing the first extra-solar planet was only published in 1988. Since this first discovery the number of extra-solar planets known at the time of writing is a remarkable 879, with over 3,000 other potential candidates. This number is likely to increase even more rapidly in the next few years, as probes orbiting above the Earth's atmosphere collect even more detailed data, and the techniques become much more sensitive.

For the most up to date information on the extra-solar planets that have been discovered, visit this Website: http://exoplanet.eu/

6.9   Measuring Double Stars

In order to keep track on all these changes, double stars are constantly measured to keep an eye on what is happening. To make more serious observations, the double star observer will be interested in measuring the following:

1. The magnitude of each member star. The brightest star within the association is considered to be the primary (or A star). The fainter star is known as the B star, etc. (sometimes referred to as *comes*). If a pair of stars are of equal brightness, the person who discovers the duality of the system chooses which star is designated as the primary.
2. The distance (separation) between the stars. This is measured in seconds of arc ("). There are a number of ways in which this measurement can be taken. The simplest way is to do this is the drift method. This is where you allow the two stars to drift across the field of view and time how long each one passes. An eyepiece with a reticule etched into it is used to measure accurately the distance between the two stars. More complicated equipment such as filar micrometers can be used. These use finely lined movable reticles, which measure the small distances between the stars. These are fairly expensive and hard to get hold of these days. A webcam can also be used to take images. The results can be analyzed using relevant software.
3. The Position Angle between each star. This is measured in relation to the primary star. It is measured as an angle moving eastwards from an imaginary line from the primary star that points towards the north pole in an anti-clockwise direction.
4. Epoch. Because the positions of the double stars relative to one another are constantly changing, the year in which the observation was made must be stated. So always make sure that your references are as up to date as possible so you have the most current data.

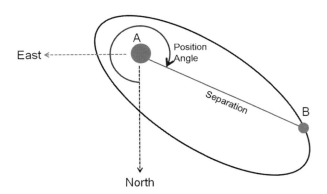

Fig. 6.3 Diagram showing the various double star measurements taken by astronomers (Courtesy of the author)

Stars orbiting further apart from one another will be easier to observe but any changes in position angle will be extremely slow. So changes may take many years to show noticeable differences.

Stars orbiting close to one another will be harder to observe due to their small apparent separation. However their position angle and separation will change much more rapidly. Stars like this will need monitoring and measuring far more frequently. Determination of the orbits of any binary system enables the masses of the component stars to be calculated. This can be extremely useful in estimating the distance to the stars.

Double stars are extremely good objects to test the quality of your equipment and your observing skills. So get out and try a number of double stars of varying difficulty to see exactly what your equipment is able to resolve. There are also quite a number of multiple stars to look out for as well, such as Sigma Orionis.

Even today astronomers are still finding new double stars, and there are many others that need other observers to go out and confirm their duplicity. There are some stars that just appear to be double. They can lie many light years from one another, but just happen to appear close together along the same line of sight. Despite not being physically associated, some of these can still be very challenging objects. Especially if they are close together and one star is much brighter than the other.

6.10   Instruments for Splitting Double Stars

The best telescope for making double star observations is one with a reasonably long focal length. This will enable you to get a higher magnification and in good seeing will enable you to maximize the separation of the two objects. A good quality refractor is the preferred choice of instrument, although a Schmidt-Cassegrain, or a Makzutov-Newtonian are extremely good alternatives. The only real true test of a telescope is to look at the image produced by your telescope. The real test of a telescopes performance is to look at the image of a star.

6.11   The Airy Disk

Stars are essentially pinpoints of light. Stars are so far away and their disks so small that even with the instruments we have today, no matter how much you magnify the star, it will never appear as anything more than a pinpoint of light. However, when you look at an image of a fairly bright star through a telescope, most of the time you will not get a pinpoint image. The light from the star will in all likelihood be boiling away to some degree. This is caused by the Earth's turbulent atmosphere above the observers head. This will somewhat spoil the image. Even if the seeing was perfectly still the resulting image will still not be as sharp as might be expected

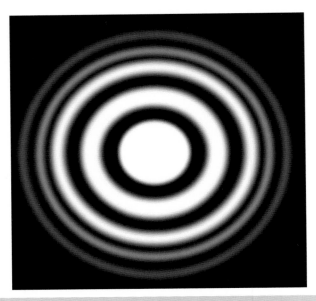

Fig. 6.4 Appearance of the airy disk in a telescope (Courtesy of the author)

and will show a small extended disk of light. This isn't due to the telescope magnifying the image, as the stars are far too far away for us to be able to do this. As well as atmospheric disturbance the nature of light and optics will result in light diffracting and spreading within the optical system to produce a much wider image than might be expected. Outside of the bright main image you should also be able to see a series of fainter concentric rings of light, where the light from that star is diffracted even further. These are known as the Airy Disk, named after George Biddle Airy. The image produced is a highly intense central peak of light and a ripple of progressively fainter rings around the outside.

Even the most perfect of optical systems will produce an Airy Disk. The size of the Airy Disk and the number of concentric rings produced by your equipment and the seeing will determine the maximum resolution of your scope.

In general terms, a smaller aperture telescope will produce a larger Airy disk.

The larger the Airy Disk produced, the further stars need to be apart from one another in order to be able to resolve them. The closest double stars would have overlapping Airy Disks and as a result they would be very difficult to separate with that setup.

A very nice Airy disk simulator is located here:

http://www.star.le.ac.uk/classroomspace/Physics/Diffraction/Airy%20Disc%20Diffraction.swf

If your Airy Disk looks like Fig. 6.5, then your telescope needs collimating.

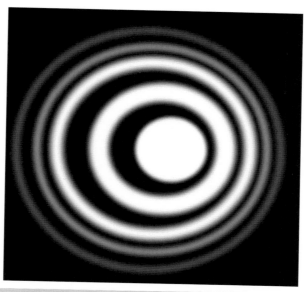

Fig. 6.5 Appearance of an airy disk in a badly collimated telescope (Courtesy of the author)

More details about The Airy Disk are found here: http://www.oldham-optical. co.uk/Airy%20Disk.htm

There are a number of diffraction limited optical systems available, but the cost of these types of optics are well out of the budget of most amateur astronomers. Much of the time our observations are limited by the turbulent atmosphere above our heads anyway. So unless you are very serious about your astronomy, and observe in a place that has many nights of exceptional seeing, using standard optics isn't going to be too much of a problem for most observers. Work within your known limits and you won't be disappointed.

6.12 Dawes Limit

This is the measurement of the closest double star that can be separated with your setup This is the maximum resolving power of your telescope. The Dawes Limit for your telescope can be easily calculated using the following formula.

$$4.65/D = r$$

Where:
D = Diameter of telescope objective lens or mirror in inches.
r = The separation of the closest star (or any other object) resolvable in arc seconds.

So an aperture of 4 in. will have a Dawes limit of:

$$4.56/4 = 1.14 \text{ arc seconds.}$$

An 8 in. telescope will have a Dawes Limit of:

$$4.56/8 = 0.57 \text{ arc seconds.}$$

It follows that the bigger the telescope, the closer the double stars that (theoretically, at least) that should be resolved. This can also be used for calculating the possibility of resolving other objects like planetary and lunar features, or the Galilean satellites, for example.

Bad atmospheric seeing on a poor night will reduce this figure significantly. So your optics may not always be performing at their best. This performance will change from night to night, and can even change hour to hour as atmospheric conditions change.

A convenient Dawes Limit Calculator is found on this Web site:

http://www.asterism.org/clubact/calc06.htm

6.13   A Couple of Challenging Double Stars

Exercise 6C: Separate Gamma Virginis

Placed about 39 light years away, this bright double star has been unresolvable by amateur astronomers for a number of years now while the stars were closer together.

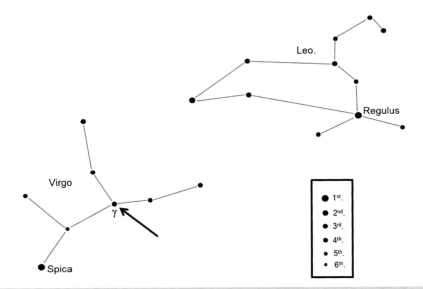

Fig. 6.6 Finder chart for gamma virginis (Courtesy of the author)

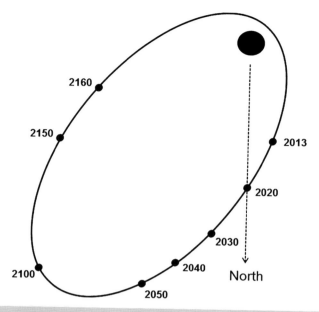

Fig. 6.7 The apparent orbit of gamma virginis as seen from earth (Courtesy of the author)

It is composed of two yellow stars both of equal brightness at magnitude 3.5 and have an orbital period of 169 years. In 2010 the two stars were at their closest to one another (periastron) separated by only 0.9″. The fact that the stars are so bright made it difficult to separate them, as the glare tended to fill the intervening space. At this time a fairly large telescope was required to split them. The distance between the two stars is now widening, so it is becoming much easier to separate them in an amateur sized scope.

Good optics and a high magnification are still required. The task will become easier over the next few years as the stars continue to separate at their widest (apastron) of 5.1″ in 2050.

Sirius

The brightest star in the whole sky has long been known to be a double star. Its presence was suspected in 1844 by Friedrich Bessel by Sirius' proper motion and was first observed directly by Alvin Clark in 1862. It has an extremely small white dwarf companion, called "The Pup" sharing its common motion in space. The primary star we are familiar with has a magnitude of −1.5, The Pup has a more diminutive magnitude of 8.5. Having a magnitude bright enough to be seen in a modest telescope you would think it would be fairly easy to spot. Unfortunately the biggest problem to viewing the companion star is Sirius itself. Its sheer brilliance tends to dazzle your field of view, flooding light around the telescope and preventing the observer from seeing it. In northern latitudes the star never rises very far

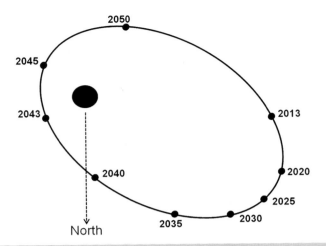

Fig. 6.8 The apparent orbit of sirius as seen from earth (Courtesy of the author)

above the horizon, so we have the added complication of a very turbulent atmo-sphere. As a result the primary star flashes and shakes violently in the high powered field of view necessary to split the companion. Fortunately we are now at a time where in its 50 year orbit the companion star is now starting to move well away from the primary. As the distance between the two stars increases over the next few years, The Pup should become easier to spot. In 1993 the two stars were at perias-tron and only 2.5″ apart so it was virtually impossible to spot the faint companion so close to the dazzling primary. In 2023 the two stars will be at apastron, so for a few years either side of this date gives the greatest opportunity. They will then slowly move towards one another. In 2044 the two stars will be reach periastron once more.

Exercise 6D: How to Catch the Pup

Try observing Sirius as the sky is just getting dark. Alternatively try and observe with the full Moon in the sky, or through thin cloud. These all help to reduce the glaring effect of the primary star, enabling you to see the dimmer companion more easily. An occulting bar in the eyepiece will definitely help in seeing The Pup. When you do spot this famous white dwarf star, you are viewing a very dense star with a diminutive size comparable to the Earth. Not bad going for an object over 8½ light years away. Keep trying over the next few years as they become wider apart.

There are many more double stars scattered around the sky that are just as compelling or challenging to observe, but we will move away from the constella-tions and stars themselves and now start to look at the nearest objects to us in our solar system.

6.14 Double and Variable Star Resources

An extremely useful list of observable double stars is available from The Belmont Society:

http://www.jouscout.com/astro/belmont/belmontd.htm

As an aside, after reading through the Belmont list, the author thinks there may be just a slight "problem" with it. It clearly describes some of the companions of these double stars as being green colored. It is quite well documented that there is no such thing as a truly green star. They just do not emit enough green light for us to see above everything else.

For a good discussion on green stars visit this Web site:

http://blogs.discovermagazine.com/badastronomy/2008/07/29/why-are-there-no-green-stars/

So why does the Belmont list describe some stars as being green? Everyone has different perception. A lot of what you see is interpretive, especially where color is concerned and everyone sees something slightly different.

Get out and look at some of the Belmont list for yourself. Try and determine the star colors for yourself and compare them after your observation to see whether or not you agree with the colors that they state in their list. It will certainly get you looking at star colors in a new light (pun definitely intended).

American Association of Variable Star Observers (AAVSO). A society dedicated to observing and monitoring the behavior of variable stars.

http://www.aavso.org/

An extremely good number of documents from the AAVSO about variable stars are available here (There is so much here that it will take some searching):

http://www.aavso.org/files/

British Astronomical Association (BAA) Variable Star Section:

http://www.britastro.org/vss/

Society for Popular Astronomy (SPA) Variable Star section:

http://www.popastro.com/sections/vs.htm

The Washington Double Star Catalogue (WDSC) is the world's principal database of astrometric doubles. This catalogue is frequently used on planetarium software. The full database resides here:

http://ad.usno.navy.mil/wds/wds.html

Double and Variable Stars from Sky & Telescope:

http://www.astronomy.com/en/News-Observing/Urban%20Skies/2006/12/Fun%20with%20double%20and%20variable%20stars.aspx

Observing Variable Stars. Gerry A. Good. Springer. 2002.

Variable Stars and How To Observe Them. Arne E Henden. Springer. 2010.

Understanding Variable Stars. John R. Percy. Cambridge University Press. 2011.

Double and Variable Stars from Sky & Telescope:

http://www.astronomy.com/en/News-Observing/Urban%20Skies/2006/12/Fun%20with%20double%20and%20variable%20stars.aspx

Double Stars for Small Telescopes: More than 2,100 Stellar Gems for Backyard Observers. Haas, S. Sky & Telescope. 2006.

Astronomical League, Double Star Program:

http://www.astroleague.org/al/obsclubs/dblstar/dblstar1.html

The Washington Double Star catalogue:

http://ad.usno.navy.mil/wds/wds.html

The Cambridge Double Star Atlas. Mullaney, J. & Tirion, W. Cambridge University Press. 2009.

Observing and Measuring Visual Double Stars. Argyle, R.W. 2004. Springer.

Astrosurf has a great Web site explain how to measure double stars with the free software Reduc, which can be used to reduce double star measurements.

http://www.astrosurf.com/hfosaf/index.htm

http://www.astrosurf.com/hfosaf/Reduc/Tutoriel.htm

StarLight Pro – Software for modeling Eclipsing Binary Stars.

http://www.midnightkite.com/index.aspx?URL=Binary

Chapter 7

The Solar System: The Sun

7.1 Observing the Sun

The Sun is the nearest star to us. At only 93 million miles distance it is the only star we can observe in great detail. However it's close proximity does bring with it one major problem. The extreme brightness and heat given off by our Sun does mean that we have to be extremely cautious and take rigorous precautions when observing it. The intensity of its radiation can permanently damage your eyes.

DO NOT OBSERVE THE SUN WITHOUT USING THE CORRECT FILTRATION. EVEN WHEN USING THESE, GREAT CARE MUST BE OBSERVED AS THE SOLAR RADIATION IS EXTREMELY DANGEROUS. INSTANT BLINDNESS CAN RESULT.

That's the Health and Safety bit out of the way. To observe the Sun safely we need a way to protect ourselves from this danger. This can be achieved by observing it indirectly. Alternatively by when using direct observing methods, you need to exclude a lot of light in order to prevent damage to your eyes.

Another thing to be aware of when observing the Sun is the effect of UV light on your skin. Make sure you keep your head covered. If you are standing out in the Sun observing for any length of time, put on high factor sun cream. Even when the weather is cold, it will be extremely easy for you to burn your skin after a few hours observing.

D. Eagle, *From Casual Stargazer to Amateur Astronomer: How to Advance to the Next Level*, The Patrick Moore Practical Astronomy Series, DOI 10.1007/978-1-4614-8766-1_7, © Springer Science+Business Media New York 2014

7.2   Observing the Sun Indirectly

Indirect methods are definitely the safest methods of observing the Sun and are usually used in preference to direct observing.

Pinhole Projection

This is widely touted as a useful method and is indeed the safest method of observing the Sun. It involves little more than two pieces of white card. One of which has a small hole in it. You hold the card with the hole up towards the Sun, so that it produces a shadow. Some of the Sun's light passes through the hole and produce a small image. If you hold the other card in the shadow of the first card you should see the image of the sun in the middle of the shadow. The only real drawback to this method is that the image can be fairly faint and is a bit on the small side. Only very large sunspots visible on the photosphere, or a partial eclipse will be discernible on the image, so this method isn't recommended for serious observing.

Exercise 7A: Using Binoculars or a Small Telescope to Project the Sun's Image

Using a pair of binoculars you can get a much brighter and bigger image of the Sun. This will allow you to see the sunspots in a bit more detail.

Before you start projecting the Sun's image using binoculars or a small telescope you need to check a few things. Are your binoculars or telescope made of metal or plastic? If they are made of plastic, do not use them for solar projection as the heat that results inside the optical tube assembly can melt some components, or even set them alight.

Fig. 7.1 Solar projection using binoculars (Courtesy of the author)

Fig. 7.2 Solar projection through a small telescope (Courtesy of the author)

Cover one side of the binoculars or cover any other optical components not being used with a lens cap. If using a small telescope make sure that the lens cap securely fastened to the finder scope. These last two steps can prevent you setting your clothes alight when trying to concentrate on looking at the image. This is something that has almost been done even by experienced observers.

The strong heat being focused right on the eyepiece could crack the lenses within it. So as a general rule of thumb do not use a very expensive eyepiece for solar projection, just in case it gets damaged when being used. Unless you are prepared to replace it, of course.

To get an image of the Sun point your binoculars or telescope at the Sun. DO NOT ATTEMPT TO LOOK THROUGH THE TELESCOPE! Use the shadow of the instrument to try to gauge how well aligned on the Sun you are. Once you have aligned everything properly you should see a round image of the Sun appear over to one side.

Keep adjusting the position until the image is as central as possible. Hold a piece of white card where the image is formed and adjust the focus until the image is sharp.

Again, a permanent record can be taken by making a sketch or taking a picture of the projected image.

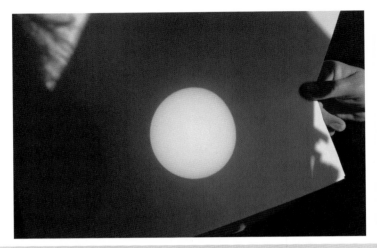

Fig. 7.3 Image of the sun projected onto a white piece of card (Courtesy of the author)

Fig. 7.4 Typical image acquired by solar projection (Courtesy of the author)

7.3 Observing the Sun Directly

This is much more dangerous than observing the image indirectly, but the view can be much more detailed, so it is worth the effort and risk, as long as you do everything to reduce that risk to a minimum. The amount of light getting into the telescope needs to be reduced significantly before you can observe safely. Always use recommended filters and do not take any chances. Your eyesight is far too important.

If you have a small telescope you can obtain a white light filter at nominal cost to go over the objective end of the telescope. These Mylar filters cut out 99.9 % of the Suns light BEFORE it enters the scope. These can be bought either ready-made to fit certain sized scopes, or as a sheet where you can cut out just the size you need to make your own. Watch out if buying these on the Internet, as there are two types of film, one is designed for visual observing and another type for imaging. The filter designed for imaging lets in a bit more light to achieve a brighter image, so is unsafe for visual use. If you have a sheet for imaging do not use it visually even for focusing a camera.

Unfortunately there are a small number of solar filters available that screw into the barrel of an eyepiece. They are often sold with some cheap telescopes sold in the high street. These are extremely dangerous. Thankfully, these are getting rarer as time goes by, but they MUST NEVER BE USED as they are just not safe. They are held very close to the focal point of the telescope where the heat is at maximum. As a result they can get extremely hot and can crack suddenly. There would then be no filter in place and all that concentrated radiation would find its way into your eye, instantly and permanently damaging your eyesight. There is also no pain when

Fig. 7.5 A safe mylar filter securely attached to the end of a small refractor for direct viewing (Courtesy of the author)

this happens. If you have one of these, throw it away, or best still smash it with a hammer and then throw it away so no-one else can ever use it.

Once you have obtained the correct filter, before putting it on the end of your telescope, always hold it up at arms-length and look at the Sun through it. You should see the small image of the Sun through it and nothing else. If you can see bright spots of light coming through, it means it has got some small holes in the film. This will let too much light and radiation through making it unsafe for use on the telescope for direct viewing. Do this before each solar observing session. If it does develop holes, throw it away and get a new one. This really cannot be emphasized enough. You really cannot be too careful with preserving your precious eyesight and the small cost involved just might save your eyesight.

If you are happy that your filter is intact, place it carefully over the end of your telescope. Make sure that it is secure and cannot fall or blow off the end of the scope. Insulation tape wrapped around the tube can be used to secure the filter and gives a little bit more reassurance. Once you are happy all is in place and secure, pop an eyepiece in and have a look through and focus.

DON'T FORGET. NEVER LEAVE A TELESCOPE SET UP POINTING AT THE SUN. Someone might just come past and try to have a look through it.

Your first view of the Sun should reveal a white disk, which, depending on the state of the Sun at that time, may have some darker sunspot patches. As the Sun is mainly composed of Hydrogen we do not see a solid surface. This photosphere is in constant turmoil and the view is changing literally moment to moment. It never looks the same, changing over the course of a few hours.

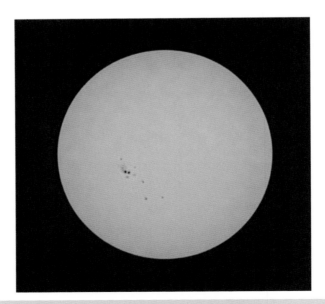

Fig. 7.6 Image of the sun taken through a white light filter (Courtesy of the author)

Exercise 7B: Draw the Full Disk of the Sun Showing Sunspots

Try and draw the Sun's full disk, noting any sunspots that might be present. Is it darker in the center of the spot? Can you see that the Sun's limb also looks a little darker towards the limb? Can you see brighter areas, called faculae, especially around Sun spots, or close to the Suns limb? How about the photosphere of the Sun, does it look a little grainy? How does changing the eyepiece change what you can see? Can you see any changes over time?

If you view the Sun over a period of a few days you will notice that the spots appear to move from left to right across the Suns disk. This is caused by the rotation of the Sun. Rotating in just over 25 days, but because of the Earth orbiting the Sun, it takes over 27 days for the same area of the Sun to be presented towards the Earth. In theory sunspots that are central on the Sun's disk on a certain day should come back to being central just over 27 days later. In fact the Sun's surface is so turbulent that unless a spot group is extremely large, they have usually all but disappeared by the time the same area of Sun rotates back into view.

The best view of the Sun will be at midday, when it is at its highest in the sky. This isn't to say that good views cannot be had a few hours either side of that. The best time of year to observe the Sun will be in the summer months when the Sun is at its highest point along the ecliptic and much higher in the sky during a large proportion of the day. From tropical regions this variation isn't quite as important as the altitude of the Sun is still quite high regardless of the time of year. The higher in the sky the Sun is, it will be well away from the murkiness of lower altitudes, which does decrease markedly the amount of detail seen. Now that we can observe the Sun safely and can draw what we see, what useful observations can we make?

7.4 Sunspots

So now we can view the Sun in comparative safety, and have enjoyed the view, let's start to do something a little more serious.

As described before the Sun's photosphere is in constant turmoil. Sunspots (also known as Active Regions) are constantly being formed and disappearing, sometimes hourly. These Active Regions are numbered using a 4-figure digit. This numbering system started on the 5th January 1972 starting at 0000 up to 9,999. When 9,999 is reached the number returns to 0000 and the numbering starts again (As it did in June 2002). The numbers are prefixed with AR for Active Region, to give a value like AR1591, which was an active region visible on the Sun in October 2012.

These relatively cooler areas of the Sun are formed where strong magnetic fields reach the top of the Sun's photosphere. Larger sunspots show two distinct areas. A much darker central region called the umbra and a lighter region around the outside, called the penumbra. They also often appear in pairs, with one having a positive charge, its companion being negatively charged.

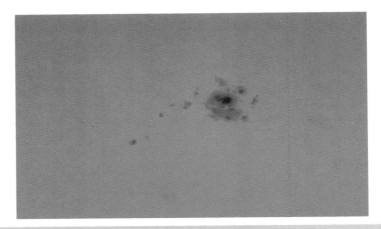

Fig. 7.7 Webcam image of a typical sunspot and group (Courtesy of the author)

The Sun goes through a cycle of sunspot activity. At maximum it is has quite a large number of active areas. The number of sunspots slowly subsides and 11 years later sunspots are much fewer and far between. After minimum Sunspot activity the first sunspots to appear are closer to the poles on each hemisphere of the Sun, gradually appearing closer to Sun's equator as the cycle progresses. This produces the characteristic Butterfly Diagram. We know that the Sun has an 11 (or 22) year cycle gradually going through these different stages. But it seems that the Sun's behavior is a bit more complicated than that. Scientists are still learning about how the Sun's dynamics "work" and are constantly learning new things about this variability. Therefore any observations made by amateurs can be of great value in adding much to our understanding of our nearest star. Even seasoned solar astronomers are constantly surprised by the ongoing behavior of our Sun.

The simplest task to start recording what the Sun is doing is to count the number of sunspots that are visible that day.

7.5 Counting Sunspots

This process really is literally as simple as it sounds. If you can view the Sun's disk safely and clearly, how many sunspots are there on its face at the time you observe? Obviously at a time of minimum activity, there will be less sunspots visible. While at maximum activity there could be quite a number.

Gathering data like this can be as simple or complicated as your patience will allow. Of course this will entirely depend on how interested you are in gathering that data and the amount of data you want to record. So let's assume you are interested in monitoring the Sun and let's take this process one step further, and really start to delve into some detail.

7.6 Counting Sunspot Groups

Sunspots regularly appear in groups. These can be small or large groups depending on the state of the Sun at the time of observation. These are usually easily identified as they occupy different parts of the Sun's disk. So if you enjoyed counting the number of spots, have a go at trying to count the number of sunspot groups. It can be a little bit more difficult to count these, so how do you distinguish between sunspot groups?

A sunspot group is normally defined as one that is over 10° away from a neighboring group on the Sun's surface. This can be very subjective as sometimes the spot groups so close together it is difficult to determine where one group ends and another begins. The key to any measurement is always to be as consistent as you can in your method for each observation.

Exercise 7C: Counting Sunspot Groups

Another measurement that can be carried out is to measure the latitude of sunspots on the Sun's disk. First make a drawing of the Sun's disk plotting the locations of the sunspot groups. A suitable template can be downloaded from here:

http://www.petermeadows.com/html/observe.html

Once your diagram has been made you will need to know the angle of tilt of the Sun as seen from Earth at the time of observation. This P-Value can be retrieved from this Java Web Page:

http://www.jgiesen.de/sunrot/index.html

Make a note of the angle of P. We can now determine the exact positions of spot groups on the Sun's surface using a Stonyhurst Disk. A number of Stonyhurst disks suitable for different size drawings can be downloaded from Peter Meadows Web page:

http://www.petermeadows.com/html/stonyhurst.html

Select a Stonyhurst Disk that is as close to the P value you noted previously. Print out the disk onto an acetate sheet and lay it over your diagram, making sure that the orientation of the disk to the north of your diagram is correct. The latitude of the spot groups that you observed can then be read directly from the disk.

Full instructions for doing so are here:

http://www.petermeadows.com/html/location.html

7.7 Classification of Sunspot Groups

As well as counting the number of sunspots, determination of the shape or morphology of the sunspot groups is also beneficial. The Sun's photosphere being such a turbulent place means that shape of a sunspot or a group is constantly changing

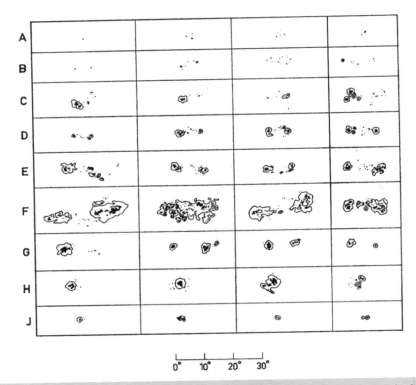

Fig. 7.8 Zurich classification diagram (Courtesy of Marco Cagnotti, http://www.specola.ch/e/disdesc.htm)

and can change from day to day. In some cases they can be clearly seen to change in just a few hours. So determining the type of spot group/s that are visible is valuable in tracking changes in spot morphology.

The Zurich Classification of Sunspot groups is the main way of classifying sunspot groups. This classification labels the groups in nine different groups:

A – One or more tiny spots the do not demonstrate bi-polarity or exhibit penumbra.
B – Two or more tiny spots that demonstrate bi-polarity, but do not exhibit penumbra.
C – Two or more spots that demonstrate bi-polarity and either the lead spot or trailing spot has a penumbra.
D – Two or more spots that demonstrate bi-polarity and the leading or trailing spot exhibits a penumbra. Occupies 10° or less of Solar longitude.
E – A group very similar to the "D" type but spreads between 10° and 15° of Solar longitude.
F – Largest and most extensive of groups. Similar to "E" type but will cover in excess of 15° of Solar longitude.

G – Decayed remnants of D, E and F groups. Exhibits a bi-polar group with penumbras.

H – Decayed remnant of C, D, E and F groups. A single spot group with penumbra. Must be larger than 2½° in diameter. The H type is occasionally accompanied by a few small spots.

J – Same as the "H" type but has a diameter less than 2½° in diameter.

More details about the Zurich classification can be found on this Web Page: http://www.specola.ch/e/disdesc.htm

The McIntosh Classification Scheme is a modified version of the Zurich system. This alternative system is starting to be used more widely. It uses a three letter classification and omitting the J category in the above classification completely.

The first letter (Z-value) is equivalent to the Zurich classification.

The second letter (p-value) indicates the appearance of the penumbra in the largest spot of the group.

The third letter (c-value) indicates the distribution of the spots within the group. This Web site explains The McIntosh Classification Scheme: http://sidc.oma.be/educational/classification.php

7.8   Hydrogen Alpha and Calcium Solar Observing

Moving things one step further, Hydrogen-alpha and Calcium light telescopes, such as those produced by Coronado are now becoming more affordable to the amateur. These use different wavelengths of light and enable the observer to see prominences arching up away from the Suns limb even without a solar eclipse. You can also see some of the fine detail and activity going constantly bubbling away on the photosphere, especially around sunspots. But these do involve spending a lot more money.

For a review of the Coronado PST, see the Cloudy Nights Web site: http://www.cloudynights.com/item.php?item_id=2832

7.9   Solar Eclipses

A total eclipse of the Sun is spectacular. Unfortunately, they do occur very infrequently from a single location. This means that unless you are very lucky you will probably need to travel quite a distance to see one in your lifetime. Totality is the one single time that the Sun can be observed without using any filtration. At the time of eclipse the light drops rapidly, especially in the last moments approaching totality.

Briefly, the following features can be observed during a solar eclipse.

The eclipse itself.

The approaching shadow of the Moon.

Timings the different phases of the eclipse.

Bailey's Beads.
The diamond Ring.
Shape and brightness of the Corona.
Prominences.

Other things to watch out for.

Birds and animals behaving strangely.
Crescent in the shadows of trees at partial eclipse.
Shadow bands.
The approach and withdrawal of the Moons shadow.
The appearance of bright stars and planets in the darkened sky.

7.10   Solar Resources

NASA Eclipse Web Site
Extremely detailed Website showing the circumstances of past and future Solar and
Lunar eclipses by Mr Eclipse, Fred Espanek.
http://eclipse.gsfc.nasa.gov/eclipse.html
Solar Influences Data Analysis Center. Solar Physics Research Department of
the Royal Observatory of Belgium. Source of the official Sunspot Number.
http://sidc.oma.be/index.php
The numbers assigned to the current (and historical) active regions visible on the
Sun's disk can be obtained from the Solar Monitor Web site:
http://www.solarmonitor.org/
Solar and Heliospheric Observatory Web Page.
http://sohowww.nascom.nasa.gov/
Solar Observing Software. Helio, Helio Viewer and Helio Creator.
http://www.petermeadows.com/html/software.html
Space Weather. http://spaceweather.com
Win Eclipse. Windows Software for predicting eclipses.
http://www.lcm.tuwien.ac.at/scs/eclipse.zip
Peter Meadows Solar Observing.
http://www.petermeadows.com/html/observe.html
How to Observe the Sun Safely. L. Macdonald. Springer-Verlag. 2012.
Solar Observing Techniques. C. Kitchin. Springer-Verlag. 2000.

Chapter 8

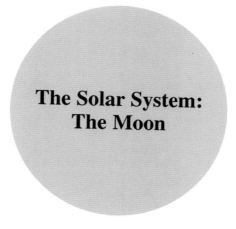

The Solar System: The Moon

Unlike most of the solar system objects, the Moon shows its face in very great detail. A wealth of features are available whatever size of telescope available to the observer.

8.1 Observing the Moon

The Moon always shows the same face towards the Earth. We all know that the Moon shows different phases according to its relative position to the Sun. Close to the Sun it shows a crescent phase, at right angles to the Sun (Quadrature) it shows a half phase. On the opposite side of the sky from the Sun it presents a full phase. The best views of lunar features are close to the areas where the Sun is rising or setting along the terminator. This is where the shadows cast by features are long and accentuated, making features seem much more rugged than they truly are. Different parts of the Moon are favored throughout the month, bringing different features into stark contrast.

This change in aspect can be significant even over the course of an hour or so. Finding interesting features, such as the Lady, The Lunar X, or V, counting tiny craterlets in the craters Plato or Clavius, or hunting down lunar domes or thin rilles running across its tortured surface will keep the lunar enthusiast extremely busy for many years to come.

At full moon, when there are very few shadows, the features look very washed out, but bright ray structures become much more prominent. At times like this look out for the craters Messier and Messier A with their Comet-shaped rays, and the very bright Aristarchus. Features closer to the limb of the Moon are frequently best

D. Eagle, *From Casual Stargazer to Amateur Astronomer: How to Advance to the Next Level*, The Patrick Moore Practical Astronomy Series, DOI 10.1007/978-1-4614-8766-1_8, © Springer Science+Business Media New York 2014

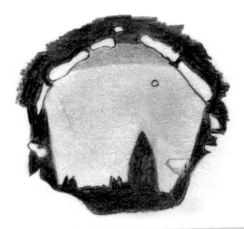

Fig. 8.1 The church shadow in ptolemaeus visible close to first quarter Moon (Courtesy of the author)

Fig. 8.2 The appearance of the lunar X just after first quarter (Courtesy of the author)

viewed a day or so either side of full Moon when they have longer shadows or are bought within view by a favorable libration.

Like most celestial objects, the Moons aspect will change throughout the month and throughout the year. The angle of the ecliptic will dictate its height above the horizon. In the evening sky, the Moon is at its highest and best viewed in the Winter and Spring sky. If observing in the morning sky, it is best viewed in the Summer

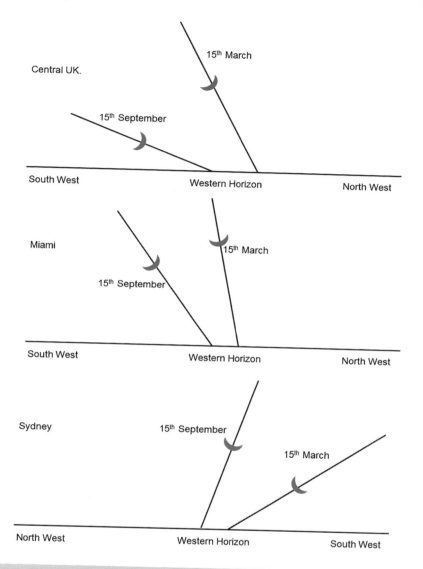

Fig. 8.3 The angle of the ecliptic in the evening skies in spring and fall (Courtesy of the author)

and Fall. This is more or less the same regardless of whether you are located north or south of the equator. The further north or south you live, the greater the change in circumstances will be seen. If you are located close to or within the tropics this effect will be much less pronounced and you should enjoy reasonable lunar observing conditions right the way through the year. This visibility depends entirely on the angle of the ecliptic to the horizon from the observer's location.

In the spring the angle of the ecliptic in the western evening sky is at a greater angle to the horizon. When favorable like this it will result in objects at a certain distance from the Sun still quite a distance above the horizon once the Sun has set. Areas closer to the equator are less affected by this angle change.

Let us now start our practice at observing features on the Moon using nothing more than your naked eye.

Exercise 8A: Draw a Naked Eye View of the Moon

Pick a time when the Moon is between half and full and easily visible. Look at the Moon and really try and see if you can reproduce what you can see onto paper. Start by marking the current phase of the Moon. Once this basic template has been created, stare at the object to try and identify some of the major features visible. You should be able to distinguish between the dark maria and the light mountainous regions in the south. Sketch those into your diagram. Can you see any of the bright rays surrounding large craters? Add those in as well. Is there anything else that grabs your attention? Look carefully and you might just be very surprised at how much detail you can make out and have added to your drawing.

You should end up with something that looks like this:

Fig. 8.4 Naked eye drawing of the Moon (Courtesy of the author)

Your drawing should be a fairly accurate representation of how much your eye can see when observing even without a telescope. This exercise has not only helped you start to map and familiarize yourself with the Moon's surface, it is also an exercise that gives you a bit of practice in drawing at the telescope. The view of the Moon with the naked eye is an image somewhat similar in size to that of some of the planets when seen magnified by the eyepiece through a telescope. So this exercise will give you some excellent practice at drawing what you can see when looking through the telescope. Using this technique of drawing at the eyepiece will enable you to reproduce what you see for future reference when observing the planets.

8.2   Common Lunar Features

Let us have a review of some of the features that can be observed through even a relatively modest telescope.

Maria – Lunar Seas

These have nothing to do with water, and are purely relatively flat features, usually large impact features that have been filled by lava flows, which occurred long after the impact. These are usually observed as being darker in nature. In fact the dark features you will have drawn in Exercise 8A will have been maria.

Mountains

The lighter areas of the Lunar surface, especially in the southern part of the observable disk is comprised of much higher features than the maria. There are a number of isolated mountain peaks that stand alone and are easily identifiable. Pico and Piton, which stand out in the Mare Imbrium, are the most prominent.

Impact Craters

Most of the cratering on the lunar surface is the result of meteoritic impacts. The size of the impact crater formed will depend on the size and speed of impact, of the body that hit the lunar surface. Most of these impacts occurred during the late heavy bombardment period, about four billion years ago, from leftover debris from the formation of the solar system. Most impact craters are therefore very old. Luckily, erosion on the Moon, being a virtually airless body is extremely slow. There is some erosion, meteorite debris and Moon tremors that do erode to some extent, but unlike on Earth, most impact carters look almost as fresh as the day they were created. There are well over 300,000 craters over 1 km in size on the near side of the lunar surface. These features vary enormously, depending entirely on their size:

Walled Plains

These are enormous and range from about 37 miles wide up to 190 miles in diameter. They characteristically have a ·flat bottom, very rugged and usually have a central mountain peak in the center. The floor is frequently covered with smaller craters, hills and rima or rilles. Clavius is an extremely good example and has lots of craterlets on its floor.

Fig. 8.5 Webcam image of the area around the walled plain clavius (Courtesy of the author)

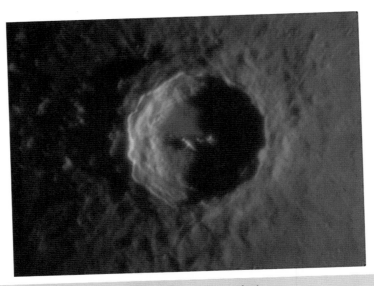

Fig. 8.6 The ringed crater copernicus (Courtesy of the author)

Ringed Mountains

Typically ranging in size from around 12 miles wide up to 62 miles across. These are extremely symmetrical craters. They display well defined, steep, circular walls, often with terracing on their inner flanks. They also show well defined central peaks. Good examples are Copernicus and Tycho. Petavius has a very distinct rima running from the outer walls towards its central peak.

Collapsed Craters

A number of craters have been disrupted in some way, or have been destroyed by subsequent bombardment. These can be extremely hard to identify.

Ghost Craters

These are large craters that have been subsequently flooded and become part of a lunar maria. Typical examples are Lamont in Mare Tranquillitatis and Stadius, close to Eratosphenes.

Volcanic Craters

Although the vast majority of craters we observe on the lunar surface are primarily impact craters, there are some small craters caused by volcanic activity. These are now being picked up on images taken by the Lunar Reconnaissance Orbiter.

An example of a fairly recent discovery in on this Web site:

http://blogs.discovermagazine.com/badastronomy/2010/08/13/ash-hole-on-the-moon/#.UWcBiJOG0rc

Craterlets

These are small craters less than 62 miles across. They are also known as crater pits. They are extremely difficult to view from Earth. The walled plain Plato has a number of craterlets that give the observer a challenge to spot on its flat floor.

Wrinkle Ridges

These are best seen under low Sun illumination close to the terminator. These features are seen as wrinkles and folds in the flat maria material which gives away the Moons turbulent volcanic past. They are a result of lave flows buckling and folding as they cool and become less fluid. Some excellent examples are seen the floor of Mare Imbrium under low illumination.

Lunar Domes

These are definitely caused by volcanic activity. There are many examples available to view on the near side of the Moon. An isolated dome, Pi, lies close to the crater Kies in Mare Nubium. A much more complex dome region Mons Rümker, over 40 miles across, lies in Oceanus Procellarum in the northern western part of the Moon.

Rima, Rilles and Clefts

These resemble cracks and splits in the lunar surface. Best observed fairly close to the terminator where they stand out much more. The central area of the Moon around Triesnecker and Hyginus is very intricate.

Using Libration to View Features Along the Limb

Another aspect of the Moon that changes is due to Libration, caused by a slight nodding of the Moon's limb towards Earth. Although we should only see 50 % of the Moon's surface, the effect of libration mean that we can view more than 59 % of the Moon's surface over time. So at certain times features close to the edge of the Moon are swung into view a little more than usual, making them a little easier to see.

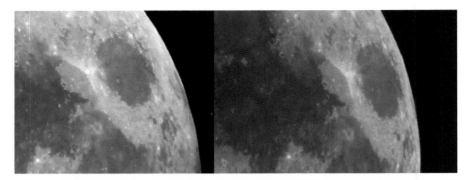

Fig. 8.7 Effect of libration on the appearance of mare crisium (Courtesy of the author)

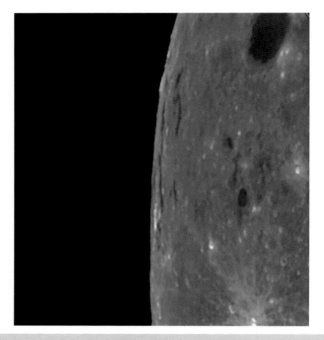

Fig. 8.8 Mare Orientale swinging into view due to libration (Courtesy of the author)

Mare Orientale

On the Western limb of the Moon there is a large impact basin called Mare Orientale (The Eastern Sea). (In 1961 the International Astronomical Union agreed that East and West on the Lunar and planet surfaces are not the same as East and West as viewed in the sky from the Earth). This colossal impact basin lies just out of view for most of the month. It is composed of a complex two concentric ringed structure of mountains and some intervening flat lava plains can be clearly seen when presented favorably. Though not all librations are the same, at certain times the effect of libration brings Mare Orientale further round the limb, so it can be viewed more easily. The more it swings into view, the more features that can be seen.

At some "favorable" librations, not much of Mare Orientale can be seen at all as sometimes it isn't presented to us quite as well. Added to this is the fact that this feature is best observed just on full Moon, or a day or so either side. Having only a small window of opportunity each year, it isn't long before years go by without ever having a chance to get even a small peek at it. Who says you soon see everything there is to see on the surface of the Moon in a very short space of time?

8.3 Transient Lunar Phenomena and Impact Flashes

As well as features on the Moon's surface, there are a couple of other possibilities to look out for. Some years ago there was a big drive to observe Transient Lunar Phenomena (TLP's). These are short lived glows, flashes of light or some other event that obscured surface details in local areas, or there may have been localized color changes on the surface. Phenomena like this have been reported by many observers over many years. This was long thought to be caused by gases escaping from the surface of the Moon. What we now know is that meteorites that hit the surface of the Moon (particularly during a very active meteor shower) do cause bright flashes as they impact the lunar surface. These have been imaged by professional and amateur astronomers alike. The bright impact crater Linné, located in Mare Serenitatis, was thought for many years to change in appearance. We now know that this was probably a trick of the light and images from orbiting spacecraft show it for what it really is, a young pristine impact crater. Observers are still reporting changes like this occurring on the Moon's surface. There is even some suggestion that these may be linked to solar activity and perhaps gas escaping from beneath the Moon's surface. It is always worthwhile looking out for some of these very subtle changes.

8.4 Lunar Eclipses

At various times in the year, the Moon can pass into the Earth's shadow to cause a Lunar Eclipse. Unlike a solar eclipse, a lunar eclipse is visible as long as the Moon is above the observer's horizon during the event. Frequently the observer will see beautiful red colors projected onto the lunar surface. This makes a wonderful sight in itself. The dedicated observer, who can drag himself away from just taking in the view, can still make other useful observations.

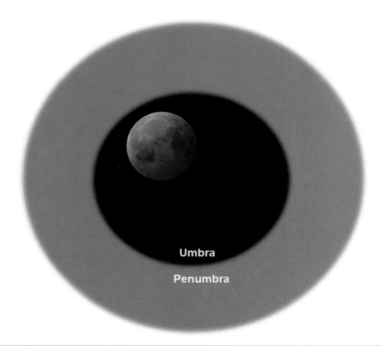

Fig. 8.9 Comparison of the sizes of the Moon and the Earth's shadow (Courtesy of the author)

The deeper the Moon goes into the Earth's shadow, the darker and redder the eclipse will appear. Eclipses further out from the central umbra will appear less dimmed. The amount of dimming and its appearance in either case will depend very much on the amount of cloud cover on the Earth during the eclipse. During a total lunar eclipse, all light reaching the Moon has to pass through the Earth's atmosphere in order to reach the Moon's surface. The light is diffracted onto the Moon by the thin layer of air covering the Earth's surface. The fewer clouds there are around the Earth's limb, as seen from the Moon, the more light that will reach the surface of the Moon. Consequently, the brighter will be the eclipse. Conversely, if there are more clouds in the Earth's atmosphere, the less light passes through and the darker the eclipse appears.

The deepness of the Moon's color during an eclipse also depends on how deep into the Earth's shadow the Moon moves during the event. The Earth's shadow has two distinct parts. The outer portion, the Penumbra, is less dark and gets darker the closer to the umbra, as more of the Suns light is obstructed by the Earth. Within the umbra no direct light from the Sun can reach the Moon's surface. An eclipse that goes straight through the middle of the umbra has the potential to be a very dark red eclipse.

8.5 The Danjon Scale

The appearance and luminosity of the Moon during an eclipse is measured using a five-point system called the Danjon Scale.

Using the Danjon Scale, you can judge the brightness of the Moon and get a measure of how much the disk has been dimmed by the Earth's shadow. This is an observation that is definitely worth reporting if you have carried it out (Table 8.1).

Table 8.1 The Danjon scale

L=0	Very dark eclipse. Moon almost invisible, especially at mid-totality
L=1	Dark Eclipse, gray or brownish in coloration. Details distinguishable only with difficulty
L=2	Deep red or rust-colored eclipse. Very dark central shadow, while outer edge of umbra is relatively bright
L=3	Brick-red eclipse. Umbral shadow usually has a bright or yellow rim
L=4	Very bright copper-red or orange eclipse. Umbral shadow has a bluish, very bright rim

Taken from NASA Eclipse Web Site:
http://eclipse.gsfc.nasa.gov/OH/Danjon.html

When the Moon is dimmed in this way, it reduces the glare from the Moon making it far easier to observe faint stars that are surrounding it. These are frequently hidden or occulted by the Moon as it moves across the sky. Occultations of background stars will be discussed a little later.

In the past, lunar eclipse observers were asked to measure as accurately as possible the time when the Earth's shadow crossed certain lunar features. This gave very accurate measurements of the relative positions between the Earth and Moon. This technique is mostly superfluous now, but could be fun to try for yourself to see just how fast the Earth's shadow does actually move across the Moon's surface.

The Sun shining down on the surface and then being cut off by the Earth's shadow during a lunar eclipse will create sudden changes in temperature on the surface of the Moon. This will inevitably produce stresses within the rocks potentially releasing gas from beneath the surface. This can obscure details or cause small changes like TLP's (as discussed previously). Therefore, it is always worthwhile keeping your eye to look out for differences on various surface features as they pass into or out of the Earth's shadow.

8.6 Lunar Occultations

The Moon moves in an easterly direction on the sky approximately its own diameter every hour, so at intervals it will pass in front other more distant celestial bodies. An occultation of a star by the limb of the Moon is the probably the most instantaneous event you can ever witness. A star, being a pinpoint of light, is instantly extinguished by the virtually airless limb of the Moon as it passes across the sky. Blink at the wrong moment and you will miss it.

As the Moon subtends an angle of about half a degree, it quite frequently moves in front of relatively bright stars situated close to the ecliptic. Many annual handbooks and almanacs list these events. You will often see the occultation predictions to name the stars by something called the ZC number. This Zodiacal Catalogue of 3,539 bright stars was published in 1940 by the Nautical Almanac Office of the US Naval Observatory.

The stars within this catalogue are mainly brighter than 8.5 magnitude and are all located within 8° of the ecliptic. Therefore, they are likely to be occulted by the Moon. Thankfully, most popular astronomical publications also use the stars popular names or the Bayer number on their list of predictions to make the stars easier to identify prior to the event.

International Occultation Timing Association (IOTA):
http://www.lunar-occultations.com/iota/iotandx.htm
http://occultations.org

Occult. Lunar Occultation Prediction software:
http://www.lunar-occultations.com/iota/occult4.htm

On rarer occasions, the Moon also occults bright planets as well. Watching Jupiter or Saturn (and their Moons) slide behind the Moon and reappear almost an hour later is wonderful to behold.

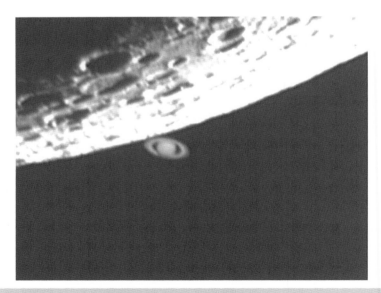

Fig. 8.10 Reappearance of Saturn after a lunar occultation in 2002 (Courtesy of the author)

Grazing Occultations

If the alignment is just right, then the occultation can become a grazing event. In this case, the star or planet just skims the very edge of the Moons northern or southern limb. Watching Jupiter or Saturn as the Moon passes just above or below, taking just the smallest chunk out of the planet is a fabulous sight. In the case of grazing occultations of stars, the light from the star can be cut off a number of times by different raised features on the Moon's surface. This quite often results in the star flashing on and off a number of times as the Moon moves past. Again, this fascinating sight really highlights the fact that the universe is not as static as it might seem at first glance. It really brings home the fact that everything in the universe is moving around and changing position the whole time.

To observe a lunar occultation, just like a solar eclipse you need to be within a narrow band of the Earth to be under the Moons shadow as it cuts off the light from the more remote object.

If you can get together with another observer who is separated from you by a pre-determined distance and you both time the occultation event, there will be notable differences between your timings. If your timings are accurate and the distance between each observer is known, then trigonometry can be used. The effect of parallax as seen by the two observers can give a very accurate estimate of the distance to the Moon.

For an example of this being carried out photographically, refer to this Web page: http://www.perseus.gr/Astro-Lunar-Parallax.htm

Lunar Occultations of Double Stars

Occasionally, the Moon will pass in front of stars that are binary or multiple in nature. This can sometimes give a view of a secondary star that cannot usually be obtained, due to the presence of the much brighter primary or the fact that the two stars are very close together. If one of the stars is occulted by the Moon first, the second star will still be visible for a few moments before the lunar limb hides that too. Even recently, cases of unsuspected and previously unresolved double stars have first been identified using this method.

8.7 Lunar Resources

The Lunar 100. Sky & Telescopes 100 Lunar features:
http://www.skyandtelescope.com/observing/objects/moon/3308811.html
Virtual Moon Atlas. Free software to show the features visible on the Moon:
http://ap-i.net/avl/en/start
Atlas of the Moon. Antonín Rükle. Excellent maps of the features of the nearside of the Moon. Now out of print. Alternatively, if you can get a second-hand copy of Moon, Mars & Venus by the same author, it is much cheaper but includes the same lunar maps.

A similar online version is available:
http://the-moon.wikispaces.com/Rukl+Index+Map

Lunar Reconnaissance Orbiter mission images can be explored in exquisite detail. Most of the lunar landing sites are visible showing evidence of the astronaut's activities.

http://target.lroc.asu.edu/da/qmap.html

Association of Lunar and Planetary Observers:

http://alpo-astronomy.org/index.htm

H.P. Wilkins' 300 in. map of the Moon.

A detailed map of the bear-side of the Lunar surface.

http://the-moon.wikispaces.com/Wilkins+300-inch+Map+Names#Wilkins 300-inch Map Names-Clickable Index to Map Sections

Chapter 9

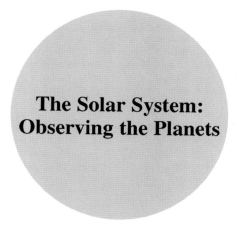

The Solar System: Observing the Planets

The visibility of the planets change throughout their orbit and depends on a number of factors, including their position in their orbit in relation to the Sun and the Earth, as well as its location along the ecliptic.

The planets come in two main types, the Inner (inferior) Planets, those that orbit closer to the Sun than the Earth – Mercury and Venus, and the Outer (superior) Planets. Those are the ones that orbit further from the Sun than the Earth: Mars, Jupiter, Saturn, Uranus & Neptune.

The visibility of the planets as they move through their orbit is determined by their position in relation to the Earth. Although all the planets orbit the Sun in the same direction, their visibility and availability for observing is dictated by their position in the solar system. Due to them orbiting closer to the Sun than the Earth, the inner planets apparitions are totally different to that of the outer planets.

9.1 Mercury & Venus: The Inner Planets

Mercury and Venus have orbits inside that of the Earth meaning that they orbit closer to the Sun. This has a major effect on how easily (or not) we can view them from Earth.

D. Eagle, *From Casual Stargazer to Amateur Astronomer: How to Advance to the Next Level*, The Patrick Moore Practical Astronomy Series, DOI 10.1007/978-1-4614-8766-1_9, © Springer Science+Business Media New York 2014

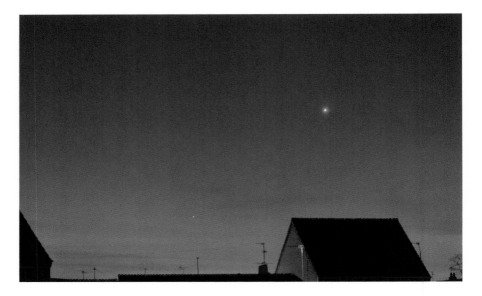

Fig. 9.1 Mercury and Venus in the early morning sky (Courtesy of the author)

9.2 Visibility of the Inner Planets

A typical observing cycle for an inner planet starts at the time when the planet is located on the far side of the Sun. This point is called superior conjunction. At this time, they will show a full disk phase. The planet will appear very small, being so much further from Earth. This, along with the fact that they are very close to the Sun's glare at this time, and on the far side of the Sun, makes them extremely difficult to observe. If you do try to find them, they will also need to be observed in daylight and will be within a few degrees of the Sun. This makes them extremely challenging to find. Do not forget to use extreme caution to protect your eyesight and equipment when observing so close to the Sun's glare.

Moving away from superior conjunction the planet moves eastwards away from the Sun. Moving further away from the Sun, they start to become visible in the western evening sky low down in the twilight just after sunset. They will be slowly approaching the Earth, so their apparent size starts to increase. Their phase changes from full to gibbous.

As they move further out of the Suns glare, their phase will change from gibbous, until they reach their furthest apparent point from the Sun. This is called eastern elongation where they present a half phase towards the observer. They will still be visible in the western sky after sunset. All the time the apparent size of the planet is increasing as the distance from Earth decreases.

This is the optimum time for observing the inner planets, when they are furthest from the Sun and visible in the darkest skies. Not all elongations are as favorable for observing the planet as one another. Different times of the year are more or less favorable for the northern and southern hemisphere. The planet then moves past eastern elongation where it will start to sport a beautiful crescent phase. At this point, their apparent size really does start to increase markedly as its distance from Earth decreases even further.

The planet will then start to move westwards back towards the Sun. At this time, it will by now, be getting much lower in the western sky after sunset. The apparent disk gets bigger all the time and the crescent getting much thinner. The closer to the Sun it becomes, the thinner the crescent. It is once again starting to court the Sun's glare as it approaches inferior conjunction. All too soon, the planet will once again be lost in the evening twilight. The planet then passes to one side of the Sun or the other reaching inferior conjunction. Like at superior conjunction, providing the planet is not too close to the Sun, they can still be observed in the daytime, if you know just where to find them. Do not forget to observe the same stringent solar observing precautions, of course. At this time, they will be visible as extremely thin crescents. On rare occasions, the planet alignment is just right so that the planet transits the solar disk. The next opportunity to observe a transit of Mercury will be on May 9th 2016. For Venus it will not occur until December 11th 2117.

Observers are quite frequently able to catch Venus in a daylight sky when it is at or very close to inferior conjunction. At this time it is an extremely thin crescent, and within a few degrees of the extremely bright Sun.

If it can be found, it presents almost a complete ring as light from the Sun is diffracted through its thick atmosphere. It is a truly wonderful sight if you can catch it. However, at risk of being boring, please ensure that light from the nearby Sun

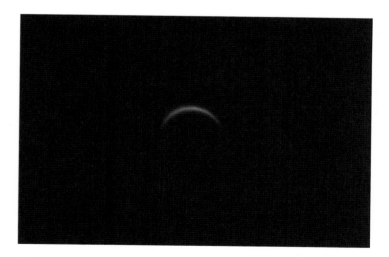

Fig. 9.2 Webcam image of Venus on the day of inferior conjunction (Courtesy of the author)

cannot enter the telescope if you do try this. You could damage your telescope, or at worse, suffer permanent eye damage if you do not take such care.

After the inner planet passes through inferior conjunction, still moving westwards it will emerge from the other side of the Sun. It will now start to appear low in the eastern morning dawn in the twilight, rising just before the Sun. The planet will move away from the bright twilight, its thin crescent phase gradually thickening. The apparent size of the planets disk decreases as the distance from Earth increases. The planet reaches its greatest distance from the Sun reaching western elongation and once again sporting a half phase, the planet will be at its best in the eastern morning sky at its furthest from the Sun.

The movement then reverses and starts resuming an eastwards track. It will then recede from Earth, the phase changing to gibbous and the apparent disk size decreasing as it moves back in towards the Sun. Moving back into the suns glare in the morning eastern twilight once more. It continues its eastwards movement as it approaches superior conjunction, where its phase will be full once again. At this time, it will only be visible in a daytime sky on the opposite side of the Sun, as before. The whole process then starts all over again. There are some other effects in the inner planets apparent movements caused by the Earths motion, but these have largely been ignored in this explanation for clarity's sake.

As both of these planets orbit closer to the Sun than the Earth, they are never seen at a great distance from the Sun. They will either be an evening object, visible for a while in the western sky after sunset, or as a morning object, visible in the eastern morning sky before Sunrise. Both planets are usually best seen when close to greatest elongation, when they are at their furthest from the Sun. To reiterate, at Western Elongation, the planets are visible in the Eastern Sky before sunrise. At Eastern Elongation, the planets are visible in the Western Sky after sunset.

9.3   Observing the Inner Planets

Mercury, being closer to the Sun and further from Earth never gets much more than 28° from the Sun. After sunset, the planet is already fairly low to the ground before the sky gets properly dark, or rises not too far ahead of the Sun in the morning sky. Venus can get up to 47° from the Sun and is much brighter, so is frequently seen higher up in a darker sky. It is therefore much easier to observe. However, its sheer brightness can cause a lot of glare in the telescope when it is observed in a dark sky.

Not all elongations are equal. The elongation of Mercury can at times be less than 18° from the Sun, 10° difference. The elongation of Venus can be less than 45° from the Sun, so with Venus there is only about a 2° difference. This of course is only the furthest and nearest elongations. Many will lie somewhere between these two figures depending on the time of year that the elongation occurs.

What does make a big difference to the visibility of both planets at elongation is the angle of the ecliptic around the time of their elongation dates. This is very similar to the effect on the changing visibility of the crescent Moon in the spring and fall as mentioned before. This angle varies depending on the latitude of the observer.

In either hemisphere, the best western elongations occur in the fall, when the planet is visible in the eastern before sunrise. The best eastern elongations occur in the spring, when the planets are visible after sunset in the western sky. Locations closer to the equator are not affected as much by these angle changes.

As stated above, both Mercury and Venus are bright enough at certain times of their apparition are observable in the daytime with a telescope. At certain times of Venus' apparition, especially in the months either side of inferior conjunction, it is even bright enough to be seen with the naked eye during the day. As it comes close to the Earth as it approaches inferior conjunction, its apparent disk size makes it just big enough to be seen sporting a wonderful thin crescent phase in binoculars. Some observers claim to have seen this with their naked eye.

9.4   Observing Features on the Inner Planets

Seeing features on the two inferior planets is extremely difficult. Mercury being so small has very little atmosphere so has no clouds obscuring its surface. Theoretically at least, you should be able to see features on its solid surface. Unfortunately as it always stays so close to the Sun it is rarely viewed in a dark sky so contrast between the subtle shadings on its surface features is extremely low. Adding to the challenge is the fact that the planet is small and its disk size always appears so tiny. As it is only usually visible low down in the sky, the turbulent atmosphere will blur any small detail that might be visible. So surface features on Mercury have and always will be a challenge visually.

Venus is covered with impenetrable bright clouds that permanently obscure its surface. So seeing surface features is impossible. Some observers have seen some elusive cloud features, usually assisted using filters. These are often described as looking like V or Y shaped features on the disk. You will need to pick a time a little while between elongation and inferior conjunction when the phase of Venus is a thick crescent. This give a good-sized disk for observing and still enough of the planets cloud tops to observe and distinguish any subtle differences.

Cusp Caps of Venus

The cusps of Venus are frequently observed to appear a bit brighter than the rest of the planet. These cusp caps may also appear to be extended a bit further round the planets disk than might be expected.

The Ashen Light of Venus

This phenomena has been long reported by some observers but is still under debate as to whether it is real or not. This is where the dark hemisphere of Venus is seen to shine faintly above the background sky. To attempt to see this, try observing Venus in a relatively dark sky when it is approaching inferior conjunction and the crescent is very slender. The planets disk is quite large at this time. Try to get the bright crescent portion of Venus hidden just outside your field of view (or use an occulting bar in the eyepiece) and see if you can spot a faint glow in the atmosphere of Venus in the dark portion of the planet. If you do manage to find it, you have joined a very select band of gifted observers who have seen this faint and extremely elusive glow.

Are any of these real effects or just tricks of the light? Whatever your feelings about this, many of the coordinators of planetary observing programs of your local or national organization would love to hear from you if you have made observations of any of these phenomena. It is always worth keeping a look out, just in case.

9.5   Mars, Jupiter, Saturn, Uranus & Neptune: The Outer Planets

The outer planets orbit the Sun further out from the Earth so their apparitions are different to those of the inner planets.

A planet that orbits the Sun beyond the Earth can only be in conjunction with the Sun when it is on the far side of the Sun. At this time the planets are usually quite small as they are the furthest from the Earth that they can be. If you could see them, they would appear to present a full phase. Unlike the inner planets the outer planets are usually much fainter and harder to see so close to the Sun, so observing them when near to conjunction is nigh on impossible.

As they move away from conjunction the planet will appear low in the eastern sky almost hidden in the morning twilight rising just before sunrise. As it moves away from the Sun it will move slowly out of the twilight, rising earlier and earlier each morning. The planet has now moved well away from the glow of dusk and gets higher in the sky each morning. Eventually they reach a point called quadrature. At this point the planet rises about midnight. The angle between the planet and the Sun is now 90°. Here the planets can show a bit of a gibbous phase. This is most obvious with Mars and Jupiter, being the two nearest outer planets to us. The best time to observe the planet at this time will be just before dawn when they will be at their highest in the south.

No matter where you are on the Earth, oppositions that occur in the summer mean that the planets are further down the ecliptic and therefore lower to the horizon. This makes features less distinct as the haze and disturbances in the atmosphere make the seeing extremely difficult. Oppositions in winter mean that the planets are much higher in the sky. Here they are much clearer to observe, being well away from most of Earth's interfering atmosphere. Observers closer to the equator are less affected by these changes in the planets declination.

As the planet carries on its orbit of the Sun, it will rise earlier each night. Eventually it will reach the opposite part of the sky from the Sun, called opposition. At this point the planet rises as the Sun sets and sets as the Sun rises. This is the best time to observe the superior planets and Minor Planets. They are then at their nearest to the Earth for that apparition and at their biggest and brightest. (There are some circumstances where the suitability of a planet will change from one opposition to another, but we will look at details of these when we discuss the individual planets). At this time details will usually be at their easiest to see. The planet is best viewed at its highest around midnight.

Once the planet moves well past opposition it starts to rise earlier each night. Soon it will only be visible in the evening sky, best viewed early to late evening. Reaching quadrature once again, the planet rises around sunset and sets at about midnight. Mars and, to a much lesser extent, Jupiter, can once again show distinct gibbous phases at this time. The planet sets earlier each evening and all too soon it starts to slip towards the western evening twilight. It will then be visible only for a short time in the western sky after sunset before setting. All too soon it will slide back into the Sun's glare and approaches conjunction with the Sun once more. Mars, moving faster across the sky takes much longer than the other outer planets to slip into the twilight. Soon the planet is lost in the evening twilight and slides towards conjunction on the far side of the Sun once again.

9.6   Retrograde Motion of the Planets

Both the inner and outer planets show some movement that is superimposed on them from due to the Earth's own motion as it moves along its own orbit. This is much more notable in the outer planets, which exhibit sweeping retrograde loops on the sky. Because the Earth moves as well it effectively overtakes them as it passes them in its faster orbit, much as a person being overtaken by a faster runner on the inner lane of a racetrack will appear to move backwards. This causes the planets to appear to slow down, stop and then reverse their direction of travel. They continue in this retrograde motion for a few months, either side of opposition, before slowing down and resuming their proper motion once again as the Earth moves away. The slightly skewed alignment of the two planets causes it to trace a

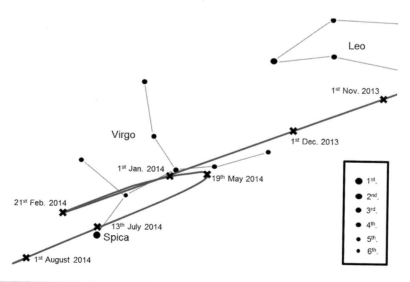

Fig. 9.3 Diagram showing Mars' retrograde loop. October 2013–August 2014 (Courtesy of the author)

loop in the sky during this time. This is most marked with the planet Mars because it is closer to Earth.

Exercise 9A: Plotting the Path of an Outer Planet on a Star Chart

Print off a map from your favorite planetarium program around the area of sky that one of the outer planets is due to come to opposition. Make sure that you leave a wide enough area of sky in the picture to take in the full retrograde loop.

Before the planet reaches opposition, go out and observe the planet. You will initially need to start observing in the morning sky. Mark its position on your sky chart on as many occasions as clear skies enable you to. Over a period of a number of weeks you should be able to see the planet slowing down and reversing its motion against the background skies.

Not only is this exercise good for showing you the motion of the outer planets, it also shows and emphasizes the way in which they appear in the morning sky, rise as the Sun sets, and set as the Sun rises at opposition and are an evening object as their apparition starts to end.

Now let us look at the individual superior planets in detail.

9.7   Observing Mars

No other planet has stirred the minds of the public more than The Red Planet. Percival Lowell was convinced that lines on the surface, which we now know to be mostly visual artifacts, were engineered water canals. The possibility of Martian life is still being researched today by the Curiosity probe on the planet surface.

Although this planet can at times be one of the closest to Earth, it can be quite frustrating to observe. It is extremely difficult to observe for most of the time, as its apparent disk size is very small as seen from Earth. It comes to opposition every

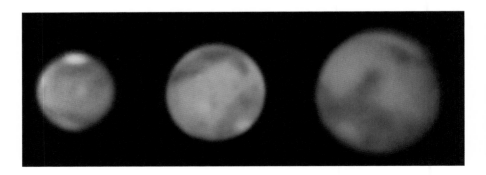

Fig. 9.4 Changes in the apparent size of Mars at different opposition distances: furthest, average and closest (Courtesy of the author)

780 days, roughly every 2 years. Therefore, there is a long wait for it between each opportunity for it to be seen at its best. Some oppositions are favorable, the 2003 event being very close at just under 35 million miles away. The opposition of 2012 saw the planet at almost double that distance, just less than 63 million miles. This difference has a huge impact on the apparent size of Mars' disk from Earth, and how much detail can be seen when viewed through a telescope.

At its closest opposition a magnification of about 150× makes the image as big as the Moon when viewed with the naked eye. Surface features on the planet will be much easier to see when the apparent disk is bigger. Conditions will improve over the next few oppositions. The planet will then be at a more southerly declination at these times, so southern observers will be favored. Mars being so small is always a challenging object. After all, Percival Lowell had one of the best refractors for observing the planet and he still thought he saw canals on its surface. Mars has a rotation period of just over 24 h. This means from an observer's point of view, if you were to go out over a series of nights at about the same time to observe the planet, it will present more or less the same features to you. Therefore, you would observe almost the same features night after night with it only appearing to rotate extremely slowly.

To see a different side of Mars, and therefore different features, you will need to extend your observing sessions or go out at different times of the night when a different part of the planet has rotated into view. Waiting a few weeks for the aspect to change can make a huge difference to what can be seen. Once Mars passes opposition its apparent size shrinks extremely rapidly and the opportunity for detailed observing is soon lost. A month or so after opposition the disk will have shrunk considerably as the distance between Mars and the Earth increases. Features frequently observed on Mars are listed below.

Dark Albedo Features

The most obvious details that can be seen on the planet's surface are areas of light and dark. These differences in reflectivity, or albedo features, are very similar to the features observed on the Moon. The dark areas being maria and the light areas more mountainous. The most distinctive dark feature on the surface of Mars is the large triangular shaped Syrtis Major. The most distinct light patch is known as Hellas. Despite being lighter in color, Hellas is in fact a low lying relatively flat impact basin.

Many other regular albedo features can be identified.

A copyrighted map by C.E Hernadez and D.M. Troiani to help identify the main albedo features is available here:

http://mars.jpl.nasa.gov/MPF/mpf/marswatch/marsnom.html

A Mars exploration map from the Mars Space Flight Facility is available here:

http://jmars.mars.asu.edu/maps/

To confirm your observations use this free app to see what Martian features were visible at the time of observation.

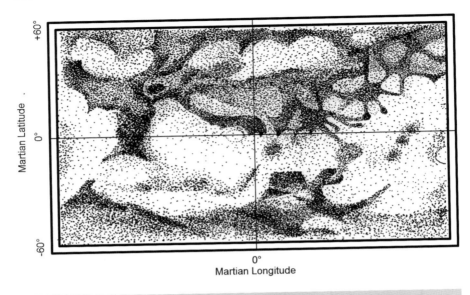

Fig. 9.5 Albedo map of the central region of Mars (Courtesy of Gita Parekh)

Mars Previewer. Free Mars Previewer simulator distributed by Sky & Telescope.
    http://www.skyandtelescope.com/resources/software/3304921.
html?page=2&c=y

Polar Caps

These are formed mainly from carbon dioxide and some water. They are visible as bright white features at both the north and south poles of the planet. The pole we can see is dependent on the tilt of the planet towards the Sun. the planet is tilted at almost 24°. The northern polar cap is only favorably tilted towards Earth when the planet is at its furthest from the Sun. Therefore, this pole is more difficult to observe. The polar caps are seen to enlarge and shrink depending on the season of the Martian year. Sometimes darker zones can be viewed around their edges. At times the southern polar cap can be seen to break up as it melts. Some detail can be seen in the cap, especially around its edges. This shows up very nicely on amateur images taken with a webcam.

Mountains/Volcanoes

Some observers have reported successfully observing some of these features. True, Mars has some of the largest shield volcanoes known, but these can never be more than a pinprick on the surface. These observations are far more likely to be due to clouds building up around these features. Nevertheless, why not have a go at looking for these features yourself. What do you see?

Atmospheric Phenomena

Clouds

Although the atmosphere of Mars is very tenuous, it is thick enough for clouds to form. It has a climate that changes according to the Martian season. Any clouds that do form seem to be transient events. They do tend build up over some of the major features on the surface, like the shield volcanoes in the Tharsis region. There appear to be different types of clouds. Whitish-blue, White, and Yellow clouds. Yellow clouds appear to precede the appearance of dust storms. Colored filters will certainly help is seeing some of these.

Dust Storms

The Martian atmosphere may be tenuous, but it is still substantial enough to lift dust particles into the atmosphere. This usually occurs during the Martian summer and is more intense when the planet is closest to the Sun. When a dust storm occurs, features, which are normally easy to observe, are obscured and difficult to make out. In July 2001 a major dust storm was observed. This obscured almost the whole surface of the planet. Minor dust storms also occur and like clouds, a colored filter will assist the observer in catching sight of them.

Seasonal Changes

As well as the changes caused by the presence of dust storms, other seasonal changes can be followed. The melting of the polar caps has already been mentioned and this can be followed very easily. Changes can occur very quickly, with the polar cap shrinking quickly over just a few weeks as the Martian summer progresses. As well as changes in the polar caps themselves, this melting also causes changes to the albedo of the surrounding terrain. So watch out for subtle changes around the polar caps. Try to observe the same features over the course of a few nights.

The darker albedo features tend to appear darker as the Martian summer approaches. Do you see this happening?

Keep a look out for anything unusual that you may not have seen before. How would you know the feature you can see is unusual? The only way to know for sure is to keep observing the planet and continually monitoring it. Only when you develop that intimate familiarity with its appearance will you be able to identify when something different is occurring.

Phobos and Deimos

The two satellites of Mars are probably minor planets that have been captured by Mars' gravitational field. When Mars is at a close opposition the satellites shine at magnitude 11 and 12. Therefore, they sound as though they should be easy to observe in a 4 in. telescope. Nevertheless, they do turn out to be challenging to the more casual observer. The reason they are so hard to observe is that they stay very close to Mars itself, so they are usually overwhelmed by the planets glare. To stand any chance of seeing these elusive objects, you will need to pick a time when either of the satellites are at greatest elongation from the planet. They will then be as far away from the glare of the planet as possible. Use of an occulting bar in the eyepiece to block out the light from Mars itself and a high power will also help. Look very carefully to see if you can see one or two faint stars close to the planet. If using a Newtonian telescope, make sure that the faint satellite images are not hidden behind a diffraction spike. Rotate the telescope if needs be. If you do spot them, can you be sure that they are the satellites and not faint background stars?

9.8   Observing Jupiter

For observers in the northern hemisphere Jupiter is currently extremely high up on the ecliptic and reaches opposition in the northern winter. This means that for a few years it will be a fine target for observers further north. For southern hemisphere observers, it reaches opposition in their summer and will remain quite low in the sky and be a bit more difficult for them to observe.

Jupiter is the largest planet in the solar system. It is a huge gas giant, so from an observer's point of view we can never view a solid surface. We see the top-most layer of clouds lit up by the distant Sun. The clouds are in constant turmoil. The most obvious features seen on the planets disk are a number of light and dark parallel lines. These are known as belts and zones.

The belts are the dark bands of clouds running across the planet. The zones are the brighter cloud belts running parallel and between the belts.. The clouds that comprise the belts are higher up in Jupiter's atmosphere and move in the opposite direction to the clouds within zones. The difference in color is caused by the turbulent atmosphere bringing different substances up from within the planet. The constant

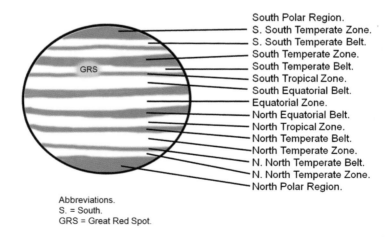

Fig. 9.6 Belts and zones visible on Jupiter and Saturn's cloud belts (Courtesy of the author)

turmoil means that the face of Jupiter is forever changing. One or two of the more permanent bands completely disappear at times. The south equatorial band is just recovering from such an event.

As the planet is gaseous different latitudes of the planet rotate at different rates. The equatorial region rotates about 5 min quicker than System II. Jupiter comes to opposition every 13 months. Being so huge its apparent disk reaches a maximum size of 50 s of arc. Even at its furthest from Earth the apparent disk is still about 30 s of arc. This means that features are readily observable throughout most of each apparition, even in relatively small scopes. The planet rotates rapidly, in less than 10 h. This causes the disk of Jupiter to squash itself into a slight oblate. Have you ever noticed this through your telescope? If you haven't, look again.

This rapid rotation does mean that features on the planet disappear rapidly over the planets limb, while new features are constantly appearing over the opposite limb. This rotation becomes apparent after only 10 min of observing. This does mean that over the course of a few hours a large portion of the planets whole disk can be observed, if you sit and wait for new features to appear.

9.9   Jupiter's Red Spot

The Great Red Spot is a huge storm that has been visible within Jupiter's South Equatorial cloud belt since 1660. It used to be a very distinct red, but has now faded to a salmon pink color. The Red Spot Hollow that the spot sits in is usually spotted before the spot itself when using a small telescope. Considering this storm is almost three times the size of the Earth, it will take a long time for a storm that size to stop revolving. As the Red Spot is located in Zone II, usually nestled within the southern equatorial band, it takes 9 h, 55 min and 41 s to rotate around the planet when at this latitude. Nevertheless, the spot does move in latitude a small amount, so this figure can change slightly. Observers from the British Astronomical Association have measured the rotation of the Red Spot over time. Their observations seems to indicate that it is now starting to rotate a little bit faster. This may suggest that the Great red Spot is starting to get smaller.

The Red spot is only visible from Earth at certain times when it is presented towards Earth. Transit times, when it can be viewed centrally on the disk, can be predicted using the Sky & Telescopes Red Spot Transit Time Calculator.

http://www.skyandtelescope.com/observing/objects/javascript/3304091.html

Many other smaller spots are seen on the cloud belts, which can come and go frequently. These are of varying hues and colors, many of them being white so keep a look out for new features. A smaller red spot located close to its bigger "brother", dubbed "Red Spot Jnr." was visible during 2012 and 2013 and seems to be a longer lasting spot.

Jupiter can be viewed in daylight, but it is a bit harder to find than Venus, as it is slightly fainter. If the Moon is close by, this can be used as an aid to locate it, focus the telescope and hop onto the planet's position. Jupiter, being so bright produces a lot of glare. This tends to scatter within the telescope, reducing the contrast and the amount of detail the observer can see on its surface. This can be overcome by observing the planet in the twilight. The lighter sky background takes out a lot of the light scattering that causes problems in a much darker sky. Features appear to have more contrast when observed at these times, making them easier to view.

Colors of Jupiter's features appear mainly white and gray through the telescope. The images taken by spacecraft and amateur webcam images are usually somewhat color enhanced. Some very elusive colors can be seen visually, mainly browns and yellows, and at times slight tinges of blue. The best way to determine the colors of the features is to use colored filters that will enhance the appearance of certain features.

9.10   Other Jovian Features

Festoons
These are seen as dark material streaming from one belt towards another. In some cases they arc round and rejoin their original belt.

Barges

These appear as darker extended objects superimposed on top of the other belts and zones.

Darkening of the Planet's Limb
The apparent brightness of Jupiter's disk is darker at the edges than the center of the disk. This effect is caused by absorbance and the scattering away of light by the atmosphere of Jupiter's atmosphere towards the edges. The brightness of the limb is also not consistent. When Jupiter is approaching opposition the preceding limb is a little brighter. At opposition, both limbs are equally bright. After opposition the following limb is brighter.

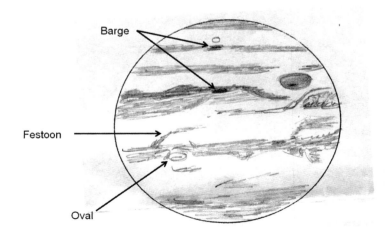

Fig. 9.7 Other features that can be observed on Jupiter (Courtesy of the author)

Impact Scars

In July 1994 Comet Shoemaker-Levy 9 ploughed into Jupiter's thick atmosphere. This caused huge scars on the cloud surface, which were easily visible in amateur telescopes. Luckily, on this occasion we managed to observe the comet before it impacted the planet, so we could predict this event. Since this has happened observers have been looking for similar events. Although none has been anywhere near as spectacular as the 1994 event, a number of smaller impacts have been observed. Two small impact scars were observed by amateurs in 2010. In some cases bright flashes are seen as a small object hits the atmosphere. In these cases the object has been too small to leave any traces of its demise once the initial flash has occurred so no scars are visible after the event.

If an impact has occurred, the longitude of the planets disk that received the impact will move rapidly out of view. If the impact zone rotates back into view when Jupiter is above your horizon and easily visible, you could be one of the first people to see the impact scar as the planets rotation reveals it to us.

The zone or belt in which it occurs and the longitudes are normally published in alert bulletins issued by many of the observing organizations mentioned in this book. Using this information, a free piece of software WinJupos, can be used to predict when that feature will be pointed towards the Earth. You can then work out if Jupiter is visible from your location at this time. It can also be used to predict the visibility of features on Saturn and the Sun as well.

WinJupos Software. Free software for calculating the position of features on planets and the Sun:

http://www.grischa-hahn.homepage.t-online.de/astro/winjupos/index.htm

9.11 Jupiter's Moons

The four Galilean moons, Io, Europa, Ganymede and Callisto, were discovered by Galileo when he used a telescope on the planet. They have magnitudes ranging from 4.6 to 5.6. These makes them easy targets and are visible even in a pair of binoculars. Some observers have reported seeing some of them without optical aid. This is theoretically possible, but you will have to find something to block the light of Jupiter from your view to stand any chance of seeing the moon so close to the glare of the bright planet. The moons positions change constantly as they orbit in the planet's equatorial plane. This means that they move from side to side of the planet passing alternately in front of and behind the planet.

Satellite Transits

When a satellite moves across the face of Jupiter, it is said to be in transit across Jupiter's disk. The satellites are much easier to see when they are closer to the limb of Jupiter. They are then in front of the darker limb region. The contrast between the planet and the moon is then optimal, making the satellite easier to observe. When the transiting satellite is further onto the disk, the much brighter background reduces the contrast between the planet and the satellite. Satellites are therefore much harder to spot whilst in transit when they are further onto the disk of Jupiter.

Satellite Shadow Transits

As well as the satellites themselves transiting the planet, their shadows can frequently be seen moving across the planets disk as well. The shadows of the satellites are extremely dark, sharp and very distinct against the bright planet. They appear larger than the satellite itself due to the size of the shadows penumbra. They can even be viewed with a modest telescope. The shadow transit is usually associated with the transit of the satellite causing the shadow. However the two events may not occur at the same time. If the transit occurs before Jupiter reaches opposition, the shadow will transit before the satellite. The satellite responsible can be observed well away to one side of the planets disk. The closer to opposition the planet is, the closer the two associated transits occur. When Jupiter is at opposition, the shadow will appear fairly close to the transiting satellite that is casting the shadow on Jupiter's disk. After opposition the satellite will transit before its shadow. The further away the planet is from opposition the further apart the two transits occur.

Satellites Passing North or South of Jupiter

As the planets orbital axis is tilted to the Sun at about 3°, at certain times in Jupiter's orbit the planet is tilted at its maximum, towards or away from the Earth. This causes some of the outer satellites to pass either above or below the planet instead of transiting or moving behind the planet. If the angle is just right, then they can pass behind Jupiter's pole, undergoing a grazing occultation behind the planet.

Mutual Events of the Galilean Satellites

When Jupiter's equatorial inclination is directly pointed to Earth (the axis is at 0°) many of the satellites can become involved in mutual events as they apparently pass one another. This results in satellites occulting one another and their shadows eclipsing each other. This can be quite exciting to watch. When watching these satellites passing one another, or transiting Jupiter's disk, you really see for yourself how dynamic the solar system really is, constantly on the move. Times of these mutual events are published annually.

Resolving Ganymede and Callisto

Unlike the other Galilean moons Ganymede and Callisto are huge. There is a possibility that with the right scope and observing conditions either of these moons might "just" be resolvable as small disks. Ganymede is the largest of the two, so is potentially the easier target. Do you think this is possible? There's only one way to find out. Get out and give it a go. To stand the best chance of doing this pick a night with steady seeing, with Jupiter close to opposition so it is at its closest to Earth. Choose a time when the planet is as high in the sky as possible (usually close to midnight around opposition). Identify Ganymede and center the satellite in the field of view. Pump up the magnification using the smallest focal length eyepiece you have. Use a Barlow lens as well if the seeing is good enough. Does it really look a lot less like a point of light? How about some subtle surface features, can you see any? You might just be very pleasantly surprised.

To make the task a little easier, try and find a time when Ganymede or Callisto are passing fairly close to Io or Europa. With the different sized satellites in the same field of view, the difference in size, using a suitable magnification, should be instantly obvious.

Jupiter with No Visible Satellites

Surprisingly this is an extremely rare event. This occurs when all the Galilean satellites are either hidden behind the planet, in transit across the face of the planet or eclipsed within the shadow of Jupiter. The last time this happened was in 2009 and was visible from most of the mainland United States. The next chance to capture Jupiter without any visible Galilean moons won't occur until 2019.

Calculate the positions of Jupiter's Galilean Satellites using this Sky & telescope Javascript facility:

http://www.skyandtelescope.com/observing/objects/planets/3307071.html#

Other Jovian Satellites

Apart from the Galilean Satellites there are many other satellites orbiting Jupiter. These are all extremely small and faint, the brightest being Amalthea which only reaches magnitude 14 at a favorable opposition. This makes them notoriously difficult for the amateur astronomer to view. A larger telescope will not help as the increased light gathering area causes even more glare from Jupiter.

Jupiter has a wealth of fine detail to take in. Its weather system and moons are undergoing constant change. This planet and its satellites really do reward the observer who is patient enough to take the time to sit down and properly take in the view.

Jupiter and How to Observe It. John W. McAnally. Springer. 2007.

The Giant Planet Jupiter. John H. Rogers. Cambridge University Press. 2009.

British Astronomical Society Jupiter Section.

http://www.britastro.org/jupiter/

Yahoo ALPO Jupiter Observers Group:

http://tech.groups.yahoo.com/group/ALPO_Jupiter/

9.12   Observing Saturn

Many people's first glance at Saturn takes their breath away. The Lord of the Rings is a beautiful sight through any telescope as the rings are stunning.

At the moment Saturn is sliding down the ecliptic into the southern sky. It currently reaches opposition in the northern spring and over the next few years will slip into the summer. Its visibility will therefore favor southern observers for the next few years, while northern hemisphere observers will need to resign themselves to the fact that details on Saturn and its beautiful ring system will become start to become harder to view in great detail.

Saturn can be viewed in daylight, but it is much harder to find than Jupiter, as it is quite a bit fainter. If the Moon is fairly close by, this can be used to orientate yourself, focus the telescope and hop to the planet. However, it really is a real challenge to spot in a bright sky, even using a telescope.

9.13   Saturn's Ring System

During Saturn's 29 year orbital period, the tilt of the planet towards Earth varies by 26°. When the inclination is zero, the rings are seen edge-on. It is difficult to observe details in the rings at these times. There are a number of events that occur around this time. The planet nodding slightly so that quite often the rings are presented edge-on twice, with the rings opening up slightly in-between these two edge-on events. This is mainly caused by the movement of the Earth. The rings were last edge-on in 2009, when it was on the other side of the Sun. The next edge on ring will be presented in 2025 but once again the planet will be fairly close to the Sun. The next best opportunity for us to view the rings edge-on in a dark sky won't be until 2039, so we have a fairly long wait to observe this properly.

The rings are divided into six main parts, lettered A – G.

A first glance through a telescope will reveal the two main rings. The brightest inner ring is known as the B ring, the outermost the A ring. Between the A and B rings is a visible gap called The Cassini Division. There are fewer ring particles orbiting at this distance from Saturn, so it is seen as a distinct dark division between the two main rings, visible even in quite a small telescope. Much harder to spot are hazy shadings visible on the A and B rings caused by the uneven distribution of particles. These are extremely challenging features to see, especially as they are relatively short-lived and cannot be predicted. Some amateur astronomers have observed what look like dark shadows called spokes within these two sections of the rings. These are caused by Saturn's magnetic field disturbing the rings, but they are short lived and their contrast with the rings themselves is extremely low, so they are extremely difficult to observe.

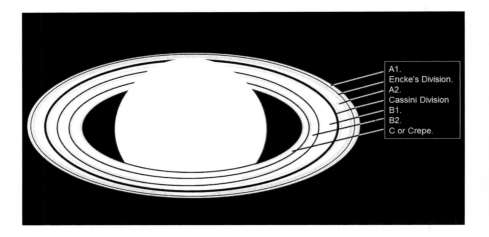

A1.
Encke's Division.
A2.
Cassini Division
B1.
B2.
C or Crepe.

Fig. 9.8 Saturn's main ring features (Courtesy of the author)

Look closely at the outer edge of the A ring. Right on the outer edge of this ring is a small division called the Encke gap. This is much smaller and less distinct than the Cassini division. It will require a larger telescope, especially under less than ideal conditions. Again it is a region where there are less particles. The B ring is also often referred in two parts, the brighter outermost named B1, the fainter B2. The other ring to look out for is the C (or Crepe) ring. This is a very faint ring that can be seen between the B2 ring and the planets disk. It is extremely tenuous, but is frequently observed in fairly small scopes, especially if the planet is close to opposition. It can be frequently spotted with an 8″ reflector. It is often revealed by its shadow that is quite often seen silhouetted as a darker shading against Saturn's much brighter disk.

As Saturn goes through opposition the particles within the rings reflect the Sun's light back more directly towards the Earth. As a result of this the rings look a lot brighter for a few days around opposition. There are many more fainter rings and divisions as revealed by robotic probes, but these are usually well out of reach of most amateurs. It is always worthwhile keeping a look out for subtle variations in the ring brightness.

Details in the rings themselves are best observed well away from edge-on presentations. The wider the rings are open, the clearer the features can be seen within the rings. The rings open and close over the orbit of the planet with two edge on and wide open presentations every 29.5 years.

9.14 Saturn's Cloud Structure

There's a bit more to Saturn than first meets the eye. Like Jupiter, Saturn is a gas giant and we observe the top of the planets clouds. Also similar is that the clouds are divided into distinct belts or zones (See Fig. 9.6). Saturn has a layer of haze hanging above the cloud tops which somewhat obscures the features below. This makes the cloud structure more challenging to observe. Careful scrutiny will reward the patient observer with these features becoming visible. The more you look, the more subtle detail and banding will become apparent.

Fig. 9.9 Changes in the presentation of Saturn's rings towards Earth (Courtesy of the author)

Like Jupiter, Saturn has a number of spots within its cloud belts. These are usually very small and fairly difficult to observe. Every few years or so a very large white spot breaks out on Saturn's disk. A spectacular white spot was observed by the amateur astronomer W. H. Hay (the actor and comedian Will Hay) in 1933. This appears to be a regular event that re-occurs every 29 years or so. The next extremely bright white spot associated with the one in 1933 isn't anticipated until 2020. It is always worth keeping a look out, as these things aren't always quite as predictable as we might like and one could appear at any time. There have been a number of white spots visible on Saturn in recent years, one that spread itself quite a way around the zone in which it resided until it faded from view. So it is always well worth keeping a look out for any changes in Saturn's cloud structure. Unlike Jupiter there have not been any reported impact scars recorded. YET!!

9.15   Saturn's Moons

Many satellites are known to orbit around Saturn. The brightest of which is Titan at only magnitude 8.3, so it is visible in most telescopes. Saturn has quite a number of satellites above 11th magnitude that can also be viewed with a fairly modest scope. How many can you identify in your setup? Don't forget that there might be a background star or two in your field of view as well, so make sure you don't confuse these with satellites.

When the planet is edge on to us and the rings virtually closed, most of Saturn's satellites appear to move more or less in a straight line. For some weeks around this time many mutual satellite events can be viewed, although they are more difficult to observe than those of Jupiter's Galilean satellites. Most of the time, Saturn's satellites appear to move above or below the planet so mutual events between them are extremely rare.

Also note that Saturn's small satellite Iapetus has long been known to be variable in brightness. It ranges from magnitude 10.2 on the west of Saturn fading to 11.9 when to the east. We now know that this is caused by a deposit of dark material on one hemisphere of its surface. When the leading side is presented to Earth we view the dark material on Saturn's eastern side. The brightest hemisphere of the satellite is viewed as it moves away from Earth on the western side of Saturn.

Next time you are looking at Saturn, take a look again and see if you can find Iapetus. Use your star brightness estimating skills learnt earlier in this book to see if you can estimate the satellites brightness. This time, there won't be any steady guide stars to compare it to. Try and repeat the exercise when the satellite is showing its opposite face and see if you too can see a difference.

You can now see that Saturn's treasures are much more subtle once you start to move beyond the main features of the rings, rewarding the patient observer with some rarely seen sights.

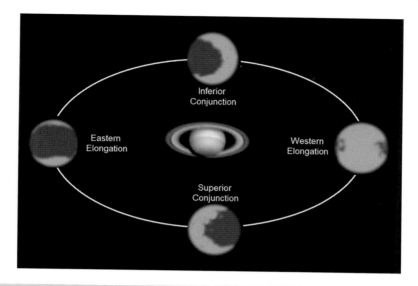

Fig. 9.10 Presentation of bright and dark features on Saturn's Moon Iapetus to explain its magnitude changes (Courtesy of the author)

9.16   Uranus and Neptune

These outer planets are extremely remote. Despite being huge planets, they are fairly faint as seen from Earth and their apparent disk size always remains small. In 2013 both planets are slowly making their way northwards along the ecliptic. This will make them much more favorable for observing from the northern hemisphere in the coming years.

At opposition the planets reach a maximum magnitude of +5.6 and +7.6 respectively. Uranus is therefore (theoretically at least) a naked eye planet. Neptune will require at least binoculars or a small telescope to identify it amongst the background stars. They move in very predictable orbits and their locations and maps of their paths are published annually, so hunting them down is fairly easy.

Through a telescope they are often disappointing. They show (at best) a very small green-blue disk. Uranus shows a disk of just over 3.5 s of arc. Neptune, being much further away but about the same size a Uranus, achieves an extremely small 2 s of arc. Seeing any surface features on either of these very small disks is nigh on impossible. Any features that might exist (were we to be able to observe it close up) are also very subtle, especially on Uranus, the nearest of the two. Even the space probe Voyager two had difficulty making out distinct features on Uranus as it shot past at close range.

9.17   Moons of Uranus and Neptune

At first viewing of these two distant planets, all we can usually see from Earth are small blue-green dots. With a bit of patience and perseverance a lot more enjoyment and a real sense of achievement can be had if you were to spot any of their faint moons. To stand any chance of spotting any of these two planets moons, you do really need to time things right. All of these satellites are extremely faint and will require tip-top optics using a fairly large aperture and crystal clear skies to reveal themselves.

The magnitude of Uranus' brightest Moons are:

Titania +13.6, Oberon +13.9, Ariel +14.2, Umbriel +14.8

Triton, the largest and brightest moon of Neptune attains a maximum magnitude of +13.6 when Neptune is at opposition. So all of these faint companions require a large telescope to stand any chance of observing them.

Spotting these elusive specks of light close to their bright primary requires a fairly large telescope, patience, perfectly collimated optics and extremely steady seeing conditions. Try using a higher power eyepiece. Sometimes the increased contrast between the background sky and the faint satellite makes them a bit easier to see.

Pick a time of year when the planet is closest to Earth (i.e. close to opposition) and a time of night when any of the brightest moons are at eastern or western elongation. The planet will be as is as high in the sky as it can get. Your chosen satellite will also be as far away from the planets glare as possible. The use of an occulting bar within the eyepiece will help to reduce the glare from the planet.

You can calculate the position of the position of the Moons using the following applets, or use your favorite planetarium program to do so.

Sky & Telescopes Uranus Moon applet.

http://www.skyandtelescope.com/observing/objects/javascript/3310476.html

Sky & Telescope's Triton Tracker applet.

http://www.skyandtelescope.com/observing/objects/javascript/13795272.html#

9.18   Planetary Occultations of Stars or Other Planets

On extremely rare occasions, just like the Moon, the planets can pass across the front of background stars. Considering the small apparent sizes of the planets, in comparison to lunar occultations, it is not surprising that this is a fairly rare event and it is even more rare for it to involve a bright star. In 1989 Saturn passed across the 6th magnitude star 28 Sagittarii. As the planet passed between the Earth and the star, the star was seen to fade and brighten a number of times as it passed behind different parts of the planets magnificent ring system. Indeed Uranus' system of faint rings was discovered using exactly the same method. Accurately timing events like this can add detail to the knowledge of the structure of Saturn's ring system. Observers in Europe also saw this star pass behind Titan the next evening.

Events like this are published in the annual handbooks, so keep a look out in current literature to see if an exciting event like this is due to take place. On even rarer occasions the planets can occult one another. Unfortunately the next event like this isn't until 22nd November 2065, when Venus passes across the face of Jupiter in daylight and only about 7° away from the Sun.

Chapter 10

The Solar System: Minor Solar System Bodies and Other Phenomena

10.1   Minor Solar System Bodies

In 2006 the International Astronomical Union has redefined the term asteroid. This class of objects is divided into two main types; Dwarf Planets and Small Solar System Bodies (SSSB's). Dwarf planets are large enough that gravitational forces produce spherical bodies. There are currently only five known examples that fit this category. There are literally thousands of SSSB's orbiting around the solar system, some potentially posing a threat to life on Earth. A major impact event seems to have occurred on Earth 65 million years ago. As there is this potential threat out there, a number of teams have been set up around the world to search for any objects, which potentially pass close to the Earth.

The brightest, Vesta reaches naked eye visibility of magnitude 5.1 when at opposition. It is therefore potentially visible to the naked eye under good conditions. Plotting its observed position on a star chart can be quite rewarding.

Exercise 10A: Plot the Position of Vesta on a Star Map

Try and identify Vesta in the night sky a few months before it reaches opposition.
    Mark Vesta's position on a printed sky map over the course of a few months.
    Finding known minor planets is challenging but fun. Nevertheless, considering the number of observatories currently operating to try and find unknown passing lumps of rock, finding a completely new discovery yourself is quite difficult today, but not totally impossible.
    There are a band of observers out there who observe recently discovered objects and take part in astrometry, measuring their positions highly accurately. From these

D. Eagle, *From Casual Stargazer to Amateur Astronomer: How to Advance to the Next Level*, The Patrick Moore Practical Astronomy Series, DOI 10.1007/978-1-4614-8766-1_10, © Springer Science+Business Media New York 2014

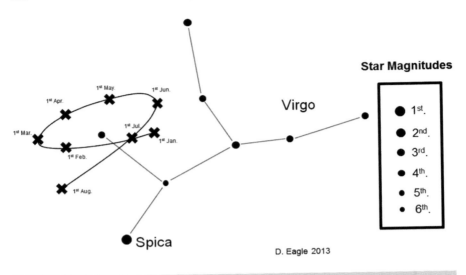

Fig. 10.1 Plot of Vesta on star map. January to August 2014 (Courtesy of the author)

observations much more refined orbits can be calculated, so that accurate ephemerides can be produced. This is invaluable in making predictions of their future orbit and positions, so observers can be informed as to where to find them and any potential threats to Earth can be quickly identified.

10.2   Pluto

Pluto is now classified as a Minor Planet. Pluto is currently around 13th or 14th magnitude and is way down in the southern sky amongst the crowded stars of Sagittarius. It has a number of much smaller and fainter companions accompanying it in its orbit. Unfortunately for northern hemisphere observers Pluto will stay quite a long way south for a few years yet, adding to the challenge of locating it. Finding this elusive planet and separating it from the many static background Milky Way stars is a real challenge.

Exercise 10B: Find Pluto

You will need at least a 10 in. telescope and dark skies. This remote solar system object can be identified by its movement from night to night. Maps of its position are published annually, so use these to identify the star pattern around it.

Find the field of view where Pluto should be. Is there a faint star that isn't plotted on your maps? It MIGHT be Pluto. The only way to be sure is to see if something has moved. Make a quick drawing of the area, plotting the position of all the stars you can see in the field of view. Go back to the same patch of sky in a day or two's time. Find the same field of view and make the same drawing again. Compare this drawing with the previous one. Can you see a "star" that has moved? If you can see that it has moved, then you have found this elusive object.

Another very extreme challenge would be to spot the presence of Pluto's largest Moon Charon. This would require a large aperture telescope as it is around 17th magnitude when close to opposition. It also never ventures very far from the much brighter Pluto. The maximum separation between the two objects is never more than 0.6 s of arc. However, if you have a large telescope to hand, or can get a look through one, why not have a go? You just never know until you try.

10.3   Get a Permanent Observatory Number

If you want to take a regular part in measuring asteroids and taking part in coordinated astrometry data gathering from a static site, you can register your observatory with the Minor Planet Center.

All details how to get your own observatory code are on these Web Pages:

http://www.minorplanetcenter.net/iau/info/Astrometry.html#HowObsCode
http://www.minorplanetcenter.net/iau/info/ObservatoryCodes.html

Generating your own observatory code will mean that the predictions made on the Web pages will be specifically calculated for your exact location. The error of doubt in the objects position will be minimal. This will maximize the chance of you being able to identify the exact position of the object and centering it exactly in your field of view.

Finding completely new objects is a very attractive proposition for the observer. Naming of astronomical objects has to be verified by the International Astronomical Union (IAU). If you do discover a completely new object, and that discovery is subsequently verified, you have the honor of naming it yourself. You can call it after yourself, a member of the family, or (almost) anything you like.

This has resulted in some familiar as well as some strange names being adopted, such as: 1772 Gagarin, 1815 Beethoven, 2309 Mr. Spock, 2602 Moore, 4147 Lennon, 4148 McCartney, 4149 Harrison and 4150 Starr.

The process of naming astronomical objects is outlined on the IAU Web page:
http://www.iau.org/public/naming/
International Occultation Timing Association (IOTA).
http://www.lunar-occultations.com/iota/iotandx.htm
http://occultations.org
Minor Planet Center Home Page.
http://www.minorplanetcenter.net/
Minor Planet Ephemeris Generator. Calculates the path and brightness of asteroids and comets as seen from a specified location.
http://www.minorplanetcenter.net/iau/MPEph/MPEph.html

10.4   Measuring the Brightness of Minor Planets

The magnitude of minor planets is not always constant. In many cases the object has differing reflective surfaces across its disk, a darker side and a brighter side. If the body is rotating, sometimes the darker side, or its smaller diameter is turned towards the observer. These changes will result in the minor planet showing distinct fluctuations in brightness. By carefully estimating the magnitude of a minor planet, details such as the distribution of light and dark patches on its surface can be determined. Any rotational period that the minor planet may have can also be deduced.

10.5   Observing Minor Planet Occultations

Like the Moon and planets, occasionally faint asteroids pass across stars and hide them from view. The stars involved show a dip in light as the asteroid, which is too faint to be seen, passes across the star. These events rarely last for more than 30 s or so. Therefore, if a 12th magnitude asteroid passes a 9th magnitude star, the brightness of the star will suddenly drop three magnitudes when the event begins and will be very noticeable. The sudden increase in brightness at the end of the event will be just as dramatic. By collating the data of the observer's results, a rough outline of the objects shape can be determined.

Observers who try observing events like this and record a non-event is not a wasted observation as it still gives us much needed information about the size of these objects and their orbits.

Online list of upcoming asteroidal occultations.

http://www.asteroidoccultation.com/

Occult Watcher. Software for the prediction of occultations of stars by asteroids.

http://www.hristopavlov.net/OccultWatcher/OccultWatcher.html

On occasion's minor planets pass bright celestial objects in the sky, or you get two minor planets passing one another. Although of course each object is many

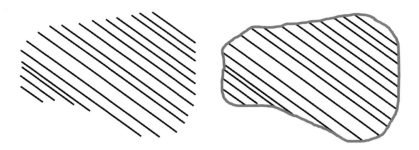

Fig. 10.2 Determining the shape of an asteroid by using different observed stellar occultations (Courtesy of the author)

millions of miles or light years away from the other, they only appear close together as seen from Earth along the line of sight. Catching two of these objects in the same field of view as they apparently pass close to one another (appulse) can be very rewarding.

10.6 Two Objects to Look Out For

2004 MN4 – This mile wide rock was discovered in June 2004. On April 13th 2029 it is due to pass the Earth at only 22,600 miles. This is well below the height at which some of the geosynchronous satellites orbit. Passing so close it will appear as a 3rd magnitude star in the sky and move about 42°/h. It will be a very fast moving easy naked eye object and observable from Europe that evening. There is no chance of it hitting the Earth as it passes.

J002E3 – This extremely small object was found orbiting the Earth in 2003. Further investigation pointed out that this unusual object may actually be the S-IVB 3rd stage of the Apollo 12 Saturn V rocket. Spectroscopic measurements revealed its surface has a distinctive match to the titanium oxide paint NASA used. Unfortunately it left Earth orbit to go back into a Solar orbit in 2003, becoming too faint to observe. Current predictions calculate that it will eventually come back to Earth, and potentially back into Earth orbit, around the year 2032. At this time it should become bright enough to be accessible to amateur astronomers once again.

10.7 Meteors and Meteor Showers

Anyone who has sat out under a dark sky cannot fail to have seen shooting stars. Fleeting flashes of light in the night sky, as particles mostly smaller than the size of a grain of sand hit our atmosphere and burn up. A lot of these objects are random events. These are called sporadic meteors and can occur at any time of the night or year. Some parts of the year are more active in meteors than others. These are known as meteor showers.

Meteor showers behave fairly predictably. They are caused by the earth passing through the orbit of a comet and ploughing through some of the debris left behind after the parent comet has passed. Not all showers are equal. If the Earth passes through the comets orbit fairly soon after the comet has passed then the shower could become a meteor storm.

The Leonids can give extremely spectacular meteor storms. In 1833 a fantastic Leonid storm was observed. We now know that every 33 years the Leonids can be spectacular when the Earth passes through a concentrated clump of debris along the orbital path of its comet of origin, P/Tempel-Tuttle. The last notable Leonid storm occurred in 2001 and resulted in extremely bright meteors that lit up the garden every few minutes, leaving ghostly trains hanging in the sky that lasted for up to 5 min. This clumping of meteoritic particles will move about and change over time.

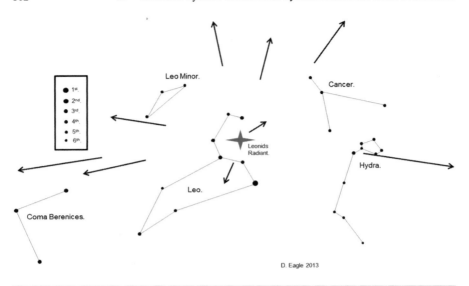

Fig. 10.3 The position of the radiant of the Leonid meteor shower (Courtesy of the author)

As time goes by new meteor showers are slowly being formed while old ones will gradually fade away. Keeping an eye on the activity of these showers tells us how these patterns are changing.

As the Earth passes through the debris the meteors are essentially moving parallel to one another. From an observer standing on the ground, meteors from a shower will appear to radiate from a single point spot on the sky. This point is known as the radiant.

The Leonid meteor shower radiant is as its name suggests located in Leo. The Lyrids from Lyra, hence the way these showers are named. Meteor showers are normally given an activity value called the Zenithal Hourly Rate (ZHR). This is the number of meteors that are predicted per hour if the radiant were located directly above your head. The further a shower meteor is from the radiant, the longer the trail will be seen in the sky.

Let us put this into context and look at the Leonid shower. It has a maximum around November the 17th each year. Leo doesn't rise in the sky until close to midnight at that time of year. So very few Leonids will be seen in the evening sky leading up to the maximum as the radiant will still be below the horizon. The meteor observer will need to wait until after midnight and start observing in the early hours of the morning. By then Leo will be fairly high up in the sky, so as a result the rate of Leonids will then be much higher. It is still not as high as if the radiant was at the zenith, so rates will still be a bit lower than the published ZHR prediction might suggest.

Look out for the presence of the Moon. A full moon will produce a large amount of light pollution. If a maxima occurs a few days around full Moon, Moonlight will hide all but the brightest meteors. The best meteor showers are when the radiant is high and the Moon is low in the sky, a very thin crescent, or best still below the horizon.

The best place to observe meteors from a shower is not right at the radiant itself but about 40° away from it.

Many more sporadic meteors are observed in the early hours of the morning. Why this occurs is better explained by trying to think about it in the following terms. At night we are looking away from the direction of the Earths orbital motion. In the morning we are looking more or less in the direction that the Earth is moving. If we use an analogy here, how many more raindrops hit your front windshield when driving down the highway? Your rear screen has far less raindrops splattered on that. By picking your moment carefully and setting the alarm clock for an early morning observing session, you really can maximize your meteor experience.

Exercise 10C: Plot a Meteor Shower on an All-Sky Chart

Pick a night when there is a prominent meteor shower. Lay on a lounger in your backyard with a copy of an all-sky chart. When you see a meteor, note its time, position, and direction of travel and length of train. Draw the trail of the meteor on your map. How bright was it? Can you estimate a magnitude? How fast was it travelling? Did it stay in one piece or break up into fragments? Did you hear anything? Some observers have reported hisses from very bright meteors. Keep plotting meteors for a number of hours. How many have you seen during the time you have been observing? Can you work out roughly how many that works out an hour? Are they all pointing towards the showers radiant, or have you observed some sporadic meteors unassociated with the shower as well? These seemingly simple observations can be very useful in monitoring the activity of meteor showers and help towards predictions of future shower activity.

It is only by observing these showers regularly and noting their activity can any changes in their behavior be identified, and the amateur observer is as always well positioned to take part in these observations.

Forthcoming Meteor Showers from The Astronomer.

http://www.theastronomer.org/forthcoming_meteors.html

The International Meteor Society collaborates amateur meteor observing.

http://www.imo.net/

10.8   Meteorites

On occasions meteorites hit the earth. The event in Russia on the 15th of March 2013 injured over 1,000 people. They were mainly hurt by broken glass from the shock wave as they were looking through the windows. Fortunately events like this are surprisingly rare. We are constantly being bombarded by debris from space that constantly rains down on us. Luckily most of the meteoritic particles are extremely small, no bigger than dust grains. These smaller meteoritic particles can be collected using nothing more than a plastic tray.

Exercise 10D: Collect Meteoritic Material in Rainwater

Take a plastic container and place it somewhere out in the open in full view of the sky. Make sure it is not under trees or any overhangs. Try and place it as high as possible in to prevent dust from the ground getting kicked in. put a few inches of water in the bottom. Leave it in the garden for at least a month. Keep topping up the water as it evaporates. The longer you leave the container out the more sample you will collect. At the end of the experiment take the container down. You should have some sediment in the bottom of the container. Not all of this will be meteoritic dust, so we now need to separate that out from everything else.

Securely wrap a magnet inside a plastic bag to make sure it is waterproof. Stir the sediment and any magnetic particles (usually meteoritic) will stick to the magnet. Transfer the dust clinging to the magnet and place the magnet in another small bowl of water. Remove the magnet from the bag and the dust will drop into the clean bowl. Keep repeating this process making sure that you stir up the sediment in your experiment bowl to ensure you collect as much sample as possible.

Once you have collected you sample, let the clean bowl dry out. Any dust left in the bottom will mainly be meteoritic in nature. Dust may be one thing but having a big lump of rock in your hand is really the best way to get a feel for the material that a meteorite is made of.

Many meteors are made of very friable materials, possibly remains of comets. These break up before hitting the ground, producing a lot of the dust that can be collected as above. The material that is hard enough to get right the way through the Earth's atmosphere and hits the ground as a meteorite are divided into three main categories:

1. Stony Meteorites. As their name suggests they are mainly composed of stony material. This groups is sub divided into a two groups, Chondrites, which have some metal composition (further sub-divided into nine sub-groups) and Achondrites, which have no metal component. This group constitutes 95 % of meteorites seen to fall.
2. Iron Meteorites. Mainly composed of iron. This group is sub-divided into 13 sub-groups depending on their chemical composition.
3. Stony-Iron Meteorites. This group is composed of 50:50 iron-nickel metal and stony material.

One of the most important meteoritic falls was witnessed by thousands of people as a greenish fireball across the East Coast of the US on October the 9th 1992. The resulting 26lb meteorite hit a red 1980 Chevy Malibu parked on a drive in Peekskill, New York. The meteorite was initially sold for $69,000 US and the wrecked car also increased dramatically in value.

So the drive to discover a meteorite is not only driven by the yearn for scientific knowledge but can also have a substantial financial value.

Meteorites fall randomly, so contrary to popular belief, meteorites can be found anywhere on the Earth's surface. Unfortunately with the Earth being comprised of over 70 % ocean. It therefore follows that 70 % of meteorites fall into the ocean and are lost for good.

Meteoritic material is very hard and resilient to weathering, so those that do make in onto the surface can last a long time sitting on the ground without harm. Very rarely is a meteor seen coming in through the atmosphere and the resulting meteorite found straight away.

10.9   Planning a Meteorite Hunt

What does a meteorite look like? Would you know a meteorite if you saw one?

Look specifically for rocks that look different to those which occur locally? Many meteorites contain iron and nickel, so a metal detector (and a magnet) will help identify those. Learning how to identify them just by eye, and being able to quickly scan the area for different rock types, is a much quicker process.

Generally meteorites are black or brown colored. They are solid and do not have pores on their surface. Most meteors are heated and partially melted as they enter the Earth's atmosphere. This results in very characteristic thumb-print indentations on their surface called regmaglypts. Meteorites are generally denser than rocks of terrestrial origin. Iron meteorites are also very heavy and magnetic.

So where do you start looking? Just about everywhere will produce samples.

The hardest task of all is to identify a meteorite from other rocks of more earthly origin. Professional collectors increase their chances of finding meteorites by looking in places where rocks stand out. They look on desert floors and in Antarctica on snow planes where any rocks will stand out like a sore thumb.

Freshly ploughed cultivated land is also quite good for finding meteorites as it contains very few naturally occurring rocks. It has even been suggested that many meteorites have been incorporated into dry-stone walls across the world collected from the walls locality when they were built many years ago. There have been over 40,000 meteorites found and cataloged to date. There must be many more out there, just waiting to be found. Many of which are probably lying in plain sight and have been over-looked by countless people, but have so far never been recognized for what they are.

In general if you find a meteorite on public land then it is freely available for the finder to keep. If you are planning to look for meteors on private land get permission from the owner of the land before hunting. Any meteors found in National Parks in the United States are considered as artifacts belonging to the federal government. As a result they cannot legally be kept by the finder. Many other countries have similar rules, so make sure that you know the rules at your search site.

Increase your chances of success by picking sites where there are fewer terrestrial rocks around such as deserts or ice fields. These sites are not always practical, especially for the amateur. If you can locate where meteorites have been seen to fall recently at the end of the track of the meteorite train they often produce "strewn fields" of debris where the object breaks up into many pieces. You can bet that many of these locations, especially if the fall occurred some time ago will have been picked-over many times before. This will reduce your success rate. So fresh fall sites offer the best chances of success, so keep an eye out for fresh meteorite falls.

These are often reported in both local and national news bulletins and the press. If you can identify the path the object has taken across a map you might be able to work out where a fresh strewn field might be located.

For more details how to find meteorites visit these sites:
http://www.novaspace.com/METEOR/Find.html
http://www.migacorp.com/meteorite_hunting_guide.htm

NASA has an All Sky Fireball Network set up with cameras that record bright meteors. http://fireballs.ndc.nasa.gov/

10.10   Observing Satellites

There are a multitude of Man-made objects orbiting around the Earth. Many of them are useful satellites, transmitting data, or observing the Earth or Universe. There is also a lot of trash metal floating around up there as well. You will have no doubt seen some of them in your astronomical career. Some of these objects are quite bright. Iridium flares are quite interesting to watch as they brightly flash in the sunlight and the ISS cannot be missed as it passes overhead with the Sun reflecting off its huge solar panels.

Most satellites in orbit around the Earth move in predictable ways. If we know which one we want to observe we know where to point our scopes.

The Web Page Heavens Above (www.heavens-above.com) generates predictions for satellites from your location enabling you to get out just before they are visible to witness their passing. If you have been observing for a while you will probably already have done this a number of times.

Let's try taking things a little deeper and set up our computers to start calculating your own predictions.

Exercise 10E: Set Up Your Computer to Calculate Your Own Satellite Predictions

Register your details at Spacetrak.org. https://www.space-track.org

This will enable you to download the latest orbital elements required for accurate predictions.

Download LXSAT software:
http://www.webtreatz.com/resources/Satellite_Tracker.zip
Join the Satellite Tracker Yahoo Group.
http://tech.groups.yahoo.com/group/satellitetracker/
LXSAT Software can be downloaded from the same location.

The update downloaded from here will enable you to download the latest orbital elements from Spacetrak.org.

After applying the latest updates to the LXSAT software, set up to download the latest orbital elements from your Spacetrak account, to keep its predictions up to date. You should now have all the software you need to make your own satellite predictions. So once you've observed a few satellites passing across and impressed your friends, what other challenges await you in observing man-made objects?

Exercise 10F: Catch the ISS as It Transits

Occasionally the ISS moves across the Moon or Suns face. Considering both objects are only half a degree across, this does happen relatively frequently from a single point on Earth. It also passes close to other bright planets at times, and its close approach can also be observed in daylight as the solar panels are now so large they reflect a lot of sunlight. To try and find out when you might be able to see one of these events, log onto the Web site www.calsky.com. Once you have registered you can put in your latitude and longitude and save it as your home location for future reference. This web page can be used to calculate all sorts of astronomical phenomena and is free to use. If you do find this web page of use, they would certainly appreciate even small donations to help with the upkeep of the website. There are a multitude of options for setting up predictions, but here's how to set things up to ONLY show Sun/Moon Transits of the ISS that shouldn't be too far from you. Log onto the Web site, selecting the Satellites page from the menu bar. When in the satellites page, select International Space Station ISS from the yellow menu at the top. Once on the specific ISS page, click the Now button to set today's date (The author has been caught out by that before). Change Select Duration to 1 Month. You could do it for longer, but the ISS orbit does change over time, so the timings will be less accurate the further in time that prediction is. Further down the page un-tick Show Satellite Passes. Tick Close Fly-bys of satellite with sun, moon planets and stars. Select 5° as the maximum angular separation from the Sun. This will restrict to events that will be a reasonable distance from you. Then tick only Sun/ Moon events. Click the GO! Button.

The page SHOULD then list a set of events. Find events that tell you that the ISS is close to the Sun or Moon. If you are in real luck, you might even get one that says that a transit is going to occur from your location. Choose an event where the Moon is as close to full as you can get it. This phase makes the most attractive images. As well as giving more of a target to capture the ISS against. Click on the link on the third box saying Centerline. This will open a view in Google Maps showing the central line of the transit. Use the slider top left to zoom right into the track. This will enable you to see the line in more detail. You only have to be about half a mile away from this central line for the ISS to miss the Moon or Sun completely as it will pass invisibly (if it is in the Earth's shadow) above or below the disk. Change Google Earth to satellite view to look in even greater detail to make sure that the site is going to be useful (and fairly safe). Clicking anywhere on that line will give you the circumstances of the transit including the predicted time.

Bear in mind that the event itself, if you do go out to observe or image the event, usually lasts much less than a second. This is especially so if the Moon or Sun is at an appreciable height above the horizon.

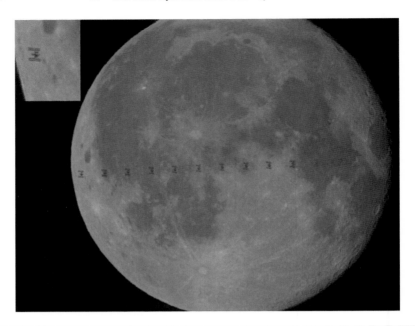

Fig. 10.4 Image of the ISS transiting across the moon (Courtesy of the author)

10.11   Observing Geosynchronous Satellites

We have all heard of geosynchronous satellites that are essential for Earth monitoring and worldwide communications. They orbit 22,236 miles above the Earth directly above the equator. This means that they orbit the Earth at the same rate that the Earth rotates. From the ground they appear to stay at the same altitude and azimuth from the observer's point of view in an arc across the sky. A person standing at the equator would see (if they could see the satellites with the naked eye) this necklace of satellites strung directly overhead going from East to West. This is commonly known as The Clarke Belt, after Arthur C. Clarke, who proposed the concept in 1945. At the Equator, this band of satellites will span 180° from East West passing right through the zenith. The further north the observer from the equator, the lower the height of this band will appear on the sky, being lower in the south. From Central UK at 52°N, this band will pass from east to west and the highest point of this arc on the meridian it will be just under 30° above the southern Horizon. From Miami, Florida, located around 26° north this band is just under 60° high in the southern sky. From Sydney in Australia, located around latitude 34° south of the equator, the arc of satellites will be around 50° high in their northern sky.

As each satellite has its own dedicated position along this band and keeps in the same altitude and azimuth position for the observer, many can be seen and kept in

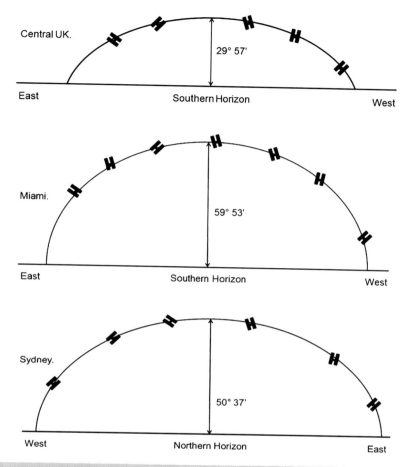

Fig. 10.5 The presentation of the Clarke Belt as seen from different latitudes (Courtesy of the author)

the field of view without having a driven scope. A Dobsonian telescope would be ideal for this purpose. So how do you start to find a geosynchronous satellite?

Observing considerations include that each satellite will vary in brightness according to the size of its solar panels, the angle of those panels to the observer and the Sun, and the amount of sunlight falling on them. Around the time of the equinoxes geosynchronous satellites are at their brightest as they reflect the Sun more directly back towards Earth as the solar panels are angled towards us.

Usually between 9th and 15th magnitude some satellites can almost attain naked-eye visibility. They will move into the Earth's shadow and be eclipsed for almost an hour and 10 min sometime during the night. Those in the East of the sky are eclipsed at the beginning of the night. Those due south around midnight. Those satellites due west are eclipsed towards the end of the night. Many of the satellites are observable on most nights of the year using an 8 in. telescope or bigger.

When they are in eclipse they do not receive any sunlight, so they are unobservable at that time. If you choose your moments right, you could even watch a satellite slowly go into eclipse. Use a telescope without a drive, or stop the drive so that, once found, your geosynchronous satellite stays within the field of view. When observed through a telescope, the satellite will appear "almost" stationary in your field of view. The background stars will be moving fairly fast behind the satellite. The satellites do jostle around a little bit due to the inclination of their orbits and eccentricities.

This Web page gives a few hints and tips on how to find one.

http://www.wotsat.com/news/article/finding-a-geostationary-satellite/9705/

For your planetarium software to be up to date, you need to download the latest orbital elements. These are found as TLE's (Two Line Elements). To get the most up to date orbital elements, use these two websites:

Celestrak.com http://celestrak.com/
Spacetrak.org. https://www.space-track.org

(Hopefully you have already registered with this Web page).

Here you can download the elements and your planetarium program should be able to handle them and display position of the satellite in real time.

Choose a fairly bright geosynchronous satellite to observe. From the US, Telstar 4 becomes reasonably bright to be easily visible. From Sydney, the Russian satellite Gorizont 29 offers the brightest target high towards the north west. From Europe the Astra satellites are fairly useful as bright targets.

As suggested above avoid periods when the satellite might be in eclipse. Use your favorite planetarium or satellite program to predict this or when your chosen target will be passing close to a reasonably bright star or deep-sky object. Don't choose too bright a star as it will dazzle your vision and may prevent you seeing the satellite as it approaches. Just before the star passes the satellite, stop the telescope drive, if you have one, and center your chosen star in the field of view and wait. You will need to keep nudging the telescope to keep the star in the field of view. Keep watching and you should be able to see a "star" that isn't moving with the rest of the sky. You have successfully bagged your first geosynchronous satellite. As your telescope is not being driven, now leave well alone. The static telescope will stay on the geostationary satellite with no effort at all and it will obediently stay in the field of view as the earth rotates. The bright star you used as a guide initially will quickly drift out of the field of view and be lost. You will then be left with a motionless "star" with all the other background stars drifting through the field of view as if caught in a current. It's quite a strange feeling to see "stars" behaving like that. Some of the satellites orbit in clusters, so if you can identify a cluster, you might be able to see a number of these looking like a small asterism of stationary stars hanging in space in the same field of view, while the other background stars silently drift past.

10.12 Observing Passing Spacecraft

Now we are going to raise the stakes even further. It is possible to observe man-made objects as they pass the Earth on a gravity assist pass. In these events, the probe may have been launched a long time previously, sent to another planet, like Venus for a gravity assisted pass. Some are often then sent back to Earth for another gravity assisted pass that gives it enough momentum to send it further into the outer solar system. An example of this was the Rosetta comet probe. This missions objective is to encounter comet 67P/ /Churyumov-Gerasimenko in 2014. Launched in 2004, it performed its first Earth Swing-by in March 2005, sending it onto Mars. It swung around Mars, picking up more momentum, passing Earth two more times, once in November 2007 and again in November 2009. The brightness of the spacecraft as it passed the Earth was well within the range of amateur telescopes. Catching up with a fast moving and accelerating object like this really does take a bit of detective work, but can be achieved.

Exercise 10G: Track Down a Spacecraft Passing Close to Earth

Your first step to identify when an event like this is likely to occur is to consult the Astronomy Magazine Spacecraft Visits Web page:
http://www.dmuller.net/space/target.php?earth
If a spacecraft is listed here and is due to make a close pass of the Earth, you then need to work out where it will be in the sky from your location as it passes. The closer you do this to the actual event, the more accurate the resulting prediction will be.
Your next port of call is the NASA HORIZONS Web site:
http://ssd.jpl.nasa.gov/horizons.cgi#top
You will need to input some details to get information accurate for your circumstances. Leave the first entry as OBSERVER.

Click Change on Target Body and enter the name of the object due to pass. You should see a selection of names if more than one object with the same name is available. Select the object you are interested in observing.
Change the Observer Location by ideally entering your latitude and longitude, or pick a listed location fairly close to you.
Change the Time Span to the date and time just before the object will make its pass, and if it's a quick pass, change the Step Size to 1 min.
Click Use Selected Settings to go back to the first screen.
Click on Table Settings and make sure that you select Skip Daylight.
Click Use Selected Settings.
Leave all the other settings as they are and click Generate Ephemeris.

It will then present a wealth of data, amongst which will be a table listing the position of the object at your chosen times and the objects brightness. Check to see if the spacecraft is bright enough to be seen in your telescope.
Is it above your horizon? Print out the details of its path in the sky and watch the maximum magnitude it will be. Is it achievable with your setup?

It's then just a case of making sure you point your telescope in the right position on the sky as it approaches the Earth and passes. Look for a star that is quickly moving when compared to other stars. How accurate are your predictions? Is it above or below the field of view you think it should be? Is it slightly ahead or behind time? Keep sweeping backwards and forwards and up and down to make sure you don't miss it. Keep coming back to the spot where the spacecraft should be at the predicted time, and keep repeating those steps. Once you do catch it moving through the field of background stars it will make your heart race and the adrenaline will start pumping. Catching an incoming spacecraft is thrilling. OK, it might only look like faint moving star, but the very thought that you have captured an object that has been around the block quite a few times, and travelled many millions of miles before it fleetingly passes by is extremely rewarding.

The Minor Planet Centre Distant Artificial Satellites Web site lists some other objects of interest: http://www.minorplanetcenter.net/iau/SpaceJunk/SpaceJunk.html

10.13   Satellite Resources

Heavens Above. Web page that can predict satellite observing opportunities from different locations. http://www.heavens-above.com/

Observing geosynchronous Satellites. http://www.satobs.org/geosats.html

Satellite Tracking Yahoo Group. Online Forum for satellite tracker software. http://tech.groups.yahoo.com/group/satellitetracker/

AMSAT's list of Satellite Tracking Software. http://www.amsat.org/amsat-new/tools/software.php

Celestrak. Source of satellite orbital elements for use in planetarium software. http://www.celestrak.com/

CalSky. Generate predictions of satellite passes from your location. http://www.calsky.com/

10.14   The Aurora

There are a wide variety of other phenomena that can be observed. Some of it not always originating from beyond the Earth. None the less, these are still interesting to observe. The Aurora Borealis and Aurora Australis are one of these. The phenomenon are caused by energetic particles reacting with gas molecules in the Earth's upper atmosphere. Although aurora can be seen many years, they are dependent on the activity of the Sun. At solar minimum auroral activity is at its lowest, or non-existent. At solar maximum there is an increased chance of seeing this phenomena.

The Earth is surrounded by a magnetic field that deflects these ionized particles away from the Earth. At both poles where the magnetic fields emerge, the particles are channeled down so that they enter the Earth's atmosphere, react with air molecules to produce the dancing lights that are seen during a display. This Auroral ring is a semi-permanent feature, with auroral activity ebbing and flowing in unison with solar activity. This is usually located around the arctic circle region, so anyone located close to this position will see aurora fairly frequently.

When there is very strong solar activity, not only are more particles sent towards Earth, increasing the possibility of a display, but the Earth's magnetic field is also distorted by the extra pressure exerted by the increased flow of particles. This distortion results in the auroral ring moving further down the globe towards the equator so that aurora can be seen much further south than normal. The stronger the activity, the further south the aurora will be seen. So what is likely to be seen if there is an auroral display?

Colors

Produced when different atmospheric gases react with the electrons. Oxygen atom above 124 km glows red. Below 124 miles an oxygen atom glows green and below 60 miles glows yellow-green. If a nitrogen atom is hit, it emits blue light. Below 62 miles nitrogen molecules glow red-purple.

For more information on aurora color see these Web sites:
http://ffden-2.phys.uaf.edu/211.fall2000.web.projects/christina%20shaw/
   AuroraColors.html
http://www.exploratorium.edu/learning_studio/auroras/difcolors.html

Curtains

The classic feature of a strong aurora. The flow of particles through the atmosphere follow the Earth's magnetic field. This produces the curtain effect. As the Earth's magnetic field changes, then the shape of the curtain will change.

Spokes

If an auroral display is particular strong and the flow of particles is coming straight towards you, perspective gives the impression of colored spokes.

Quiescent Glow

This is a faint glow of aurora close the horizon. This type of auroral activity is often overlooked or mistaken for distant man-made light pollution.

Searchlights

You may often see beams of light shining up from the northern horizon (southern horizon in the southern hemisphere) during an auroral display. These will change rapidly.

Cloud Silhouettes

As the auroral display is at least 60 miles above the Earth's surface. Most cloud features are well below this level. So if there are clouds present they will be seen as dark silhouettes superimposed against the light display.

You can get alerts or sign up for E Mail alerts for auroral activity on these Web sites:

http://auroranotify.com/alerts/
http://www.ips.gov.au/Geophysical/2/4
http://aurorawatch.lancs.ac.uk/

10.15   The Zodiacal Light

Only visible from very dark sky sites, the Zodiacal light is usually observed as a triangular shaped glow seen in the western sky a while after sunset. It is located roughly along the ecliptic and is caused by sunlight scattered off dust particles spread out in the solar system. It can in exceptionally dark circumstances be seen as a complete band along the ecliptic.

Gegenschein

This is a brighter part of the Zodiacal band of light in a direction directly opposite to the Sun. The particles in this direction have no shadows on them so appear a proportion brighter than the rest of the Zodiacal light in the same general direction of sky. It is still extremely faint, so only observed in the darkest of skies.

10.16   Atmospheric Effects

During the day and very often at night, atmospheric effects can be observed in the sky. We are probably all familiar with Rainbows, halos and Sun-dogs.

Have you heard of Moon bows and circumzenithal arcs? Most of these affects are caused by ice crystals in the Earth's upper atmosphere. But as an observer they are still extremely interesting to see.

Sun Pillars occur on a fairly regular basis. The Green Flash, which is caused by both diffraction and refraction of the Sun's light as it sets (or rises) are very rare indeed requiring ideal conditions for the event to occur.

Noctilucent Clouds

These glows towards the poles of the Earth have only been observed over the last 100 years or so. They are usually seen around mid-summer. It is thought that they are thin clouds that are high enough in Earth's atmosphere to still be illuminated by the Sun. What causes them is still a matter for debate.

For more details of some of these effects visit this Web page:

http://www.webexhibits.org/causesofcolor/13D.html

Fig. 10.6 A Lunar Halo and Jupiter (Courtesy of the author)

Chapter 11

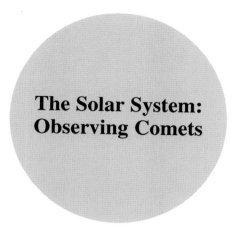

The Solar System: Observing Comets

A spectacular comet can be one of the most beautiful objects in the sky. It's gorgeous tail (or tails) can stretch quite a way across the sky. The fleeting but spectacular apparition of Comet Hyakutake (C/1996 B2) in 1996 was seen with the head close to Polaris and the tail reaching right down into Coma Berenices. A beautiful sight indeed.

11.1 Comet Behavior

While a comet is far away from the Sun, they are fairly quiet. As their orbit brings them closer to the Sun, solar radiation heats the comets surface more strongly. The closer to the Sun, the more intensive this heating becomes. The comet's nucleus contains volatiles. As these are heated, any volatiles at the surface are excited, start to boil off and are rapidly ejected from the comets nucleus by the Suns radiation. The comets apparent size and brightness will increase in proportion to this activity. An ion tail and a dust tail could also develop. Multiple tails could also form.

When the comet is at its closest to the Sun (Perihelion) the solar heating is at its maximum. This will release even more fresh volatiles further increasing the brightness and size of the comet. At these times there is potential for the comet to be ripped apart both by tidal forces (depending on how close it gets to the Sun) or the intense radiation. In some cases, the comet may break apart at or just after perihelion. If a comet does fragment, its brightness will decrease rapidly as the material disperses and the comet disappears. As the comet increases in brightness multiple tails may become visible. If it gets brighter than magnitude −1.5, then it has

D. Eagle, *From Casual Stargazer to Amateur Astronomer: How to Advance to the Next Level*, The Patrick Moore Practical Astronomy Series, DOI 10.1007/978-1-4614-8766-1_11, © Springer Science+Business Media New York 2014

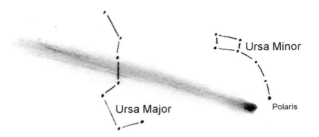

Fig. 11.1 Sketch of naked eye view of comet Hyakutake in March 1996, with tail stretching from Polaris through to Coma Berenices (Courtesy of the author)

the potential of (just) being visible in a daytime sky. In reality comets at perihelion are usually close to the Sun in the sky, so will need to be of at least magnitude −4 to be visible throughout the day and become a true "Daylight Comet".

Unfortunately bright comets like this are extremely rare. We do have a stream of fainter visitors that are visible to amateurs on a fairly regular basis. Most comets that are observed are usually fairly faint, only showing as fuzzy patches in the scope, perhaps showing a slightly blue-green tinge due to the chemicals they contain.

11.2   Comet Naming

New comets are usually named after the person that discovers them. Unfortunately with the many automated sky surveys currently in operation, a lot of new comets are identified on images they obtain. Even if a person is responsible for their identification on that image, the naming glory goes to the name of the survey that was used to take the image. It is still possible for patient amateurs to discover their own comet but it does require patient observing and diligence. It is estimated over 300 h search time is required to find a new comet visually.

Up to three independent discoverers are allowed to be used on a comets name. This is usually why you find long-winded names given to many comets, such as IRAS-Araki-Alcock. As well as the discoverer's name(s) shown above, a comet is also given a more formal name. Comet IRAS-Araki-Alcock is also known as C/1983 H1. To see where this name comes from, let us use a recent comet that was seen in 2012–2013. Comet LINEAR C/2012 K5 was discovered on images taken by the LINEAR Project (Lincoln Near-Earth Asteroid Research) in 2012. So the number indicates the year of discovery. The prefix before the number indicates the nature of the comets orbit.

The prefixes that are used are:

C/ – Non- periodic comet. The comet is in such an elliptical orbit that it will not return to the Sun.

P/ – Periodic comet. The comet has a predicted orbit and will return at some point in the future. Most of these are now documented.

X/ – Not calculated. A comet where the orbit could not be accurately measured (usually from historical records).

D/ – A comet that is thought to have broken up or been lost, or short period comets who's orbits cannot be calculated due to limited observations.

A/ – A comet that has now been reclassified as a minor planet. (Where exactly does this distinction between the two objects lie?).

A number of historical comets like Halley (1P/Halley) are also designated a P/prefix indicating their periodicity with the number showing in which order their periodic orbits were calculated. The letter after the year indicates which half of which month of the year the comet was discovered.

Letters I and Z are not used in this lettering system. A would be the first half of January, B the second half. The first half of March would be E, the first half of August would be P and on. Using this system the letter K, in our example, indicates that Comet LINEAR 2012 K5 was discovered in the first half of June 2012. The number after the letter indicates that Comet LINEAR 2012K5 was the fifth comet that was discovered during the first half of June 2012.

When a new comet is discovered its orbit needs to be determined fairly quickly. This is achieved by observation and measurement of its movement against the background sky. The direction, speed and magnitude of the comet are important measurements to take. These details can then be passed onto other observers to confirm the find.

11.3   Comet Features

Astronomers rely on many observations, plotting the position of recently found comets, to be able to calculate their orbits accurately. Plotting the movement of a comet on a star map is good practice in itself. How about estimating a comets brightness? Comets are quite difficult to estimate as they are extended objects.

Exercise 11A: Estimate the Brightness of a Comet

One of the most useful things an observer can do to a comet is to estimate its brightness. As a comet is an extended object this is a fairly involved system, and may take many times practicing to get right. If you have a comet visible, look at the comet through the telescope and look at the extent of the comets coma in the eyepiece. Move the telescope to a star you think is close to the same brightness as the comet.

Defocus the image of the star until the star appears the same diameter as the comets coma. Once the star is out of focus and image of the star looks the same size as the comet, which looks brighter through the scope, the comet or the star? If the de-focused star looked brighter, try the same exercise on a fainter star. If the star looks fainter, try on a brighter star. Keep going until you find a star that looks the same brightness as the comet when de-focused in the same way.

Once you have found a suitable candidate star look it up in a star atlas to identify exactly which star it is. Look in a star catalog, online, or in your favorite planetarium program to find out its exact magnitude. The magnitude of the star you have now identified will be the same as your estimation of the magnitude of the comets coma. How about the nucleus. Can you use the same method to get a magnitude of that?

If you do successfully estimate the brightness of a comet, please send in your observations to the relevant observing coordinators. Your observation, especially of it is of a new comet, or one approaching perihelion, will be especially useful to add to other observers observations to help in making predictions of a comets later performance more accurate.

So we now have a comets magnitude, what else can we do? Let us now look at the comet itself in more detail.

11.4   Structure of a Comet

A "typical" comet displays many of these common features:

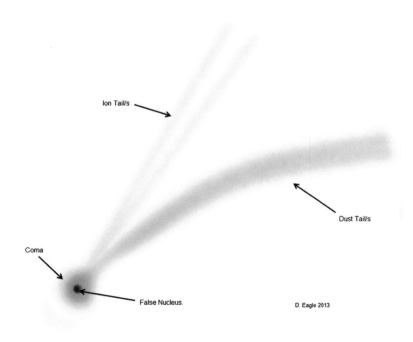

Fig. 11.2 Typical features typically observed in a comet (Courtesy of the author)

Fig. 11.3 Apparent size of comet 17P/Holmes compared to Moon in November 2007 (Courtesy of the author)

Of course there really isn't such a thing as a typical comet. The diagram only gives an indication of the type of features that might be present. Comet Holmes (17P/Holmes) was discovered in 1892. Known for previous outbursts, at its predicted return in 2007 it suddenly increased in brightness. In just a couple of days it unexpectedly flared from magnitude 17 to 2.8, making it easily visible to the naked eye. To the naked eye it had no tail to speak of (unless you did long exposure imaging from dark locations) but became extremely rounded, its actual size eventually becoming bigger than the Sun, although it was a very tenuous cloud of ejected material. This made it very exciting to watch night after night as it increased in apparent size to over half a degree in size.

The Nucleus

At the center of a comet is a small solid object. This is not visible from our viewpoint on Earth. It is just too small. In 1910 Halley's comet passed across the front of the Sun. Despite observatories in the southern hemisphere looking intently for it, no trace of it could be seen.

The solid nucleus contains a lot of icy volatiles. These coalesce when acted on by the radiation form the Sun and start to "boil off". This creates an envelope of gas around the comet. This creates a bright point of light that is full of gas and dust. This gives the appearance of a nucleus right at the heart of the comets head.

This is known as the False Nucleus, because it isn't the solid object itself that you can see. Try and estimate the brightness and size of the false nucleus.

The Coma

The more volatiles that are released, the bigger the cloud of dust and gas that surrounds the nucleus. This large shell of dust and gas around the nucleus is called the coma. The coma can potentially have different shells of gas each with a slightly different brightness. Look for any structure in this shell of gas. Are there multiple layers? Can you see any differences in brightness, or some degree of differentiation within it? Is it completely spherical, elongated or shaped in some way?

The Coma is usually the brightest part of the comet. The closer that the comet approaches the Sun, the more radiation it is subjected to and the bigger and brighter the coma becomes. The pressure of the Suns radiation then pushes the particles out to form a tail (or tails), but we will discuss these later.

11.5   Measuring the Comet's Coma

Coma Diameter Measurements

This is a measure of the bright area surrounding the comet nucleus. This is most easily measured by using the drift method. This is achieved by stopping the drive of the telescope to let the comets image drift from one side of the field of view of your eyepiece until it reaches the other edge. If you know the field of view obtained with this telescope/eyepiece combination, (You do know this by now, don't you?) then a reasonable estimated apparent size of the coma can be determined.

Degree of Condensation (DC)

This is a measurement that is frequently used to determine how concentrated or condensed the coma of the comet is. This is measured by estimating the intensity of the coma at a distance from the central nucleus.

Degree of Concentration is measured on a scale from 0–9.

0 = Diffuse coma of uniform brightness.
1 = Diffuse coma with slight brightening towards center.
2 = Diffuse coma with definite brightening towards center.
3 = Centre of coma much brighter than edges, though still diffuse.
4 = Diffuse condensation at center of coma.
5 = Condensation appears as a diffuse spot at center of coma. Moderately condensed.
6 = Condensation appears as a bright diffuse spot at center of coma.
7 = Condensation appears like a star that cannot be focused. Strongly condensed.
8 = Coma virtually invisible.
9 = Stellar or disk like in appearance.

If you think that your estimate of the comets CD is between two values, say between 4 and 5, record it as 4/.

Take some quality time to look at the comet in detail. Is the nucleus stellar-like, or is it a more extended object? Is the coma clear cut? Is the coma evenly bright, or are there patches within it that are brighter, or darker? Is the coma symmetrical, or asymmetrical? Can you see any structure within the coma?

Is there more than one Coma/Nucleus? This can often be a sign that the comets nucleus is starting to break apart. When this happens a comet can suddenly brighten dramatically as fresh volatiles are released and excited by solar radiation.

Sometimes jets, spines, fountains, rays or fans may be seen emanating from the central nucleus. Jets were particularly prominent in comet Hale-Bopp. Another feature to look for, especially in a particularly bright comet is a hydrogen corona surrounding the coma. There may also be an envelope or hood

11.6   The Comets Tail(s)

The other feature that really stands out in bright comets is the tail streaming away from the brighter coma and nucleus. It always points more-or-less in the opposite direction to the Sun, as it is the solar radiation driving it away from the comet. Therefore, the tail often precedes the comet. The comets tail can take many forms, from a single straight thin tail, to multiple curved tails with knots and other structures contained within.

If a comet has an appreciable tail, try and estimate its length. Try not to be influenced by other observer's reports. It's your estimate that matters. Your skies will be different and may give it a different appearance. There are two types of tails:

Type I
This tail is composed of ions. As these are extremely small particles they are blown away from the comet very rapidly, so these tails are usually very straight. They show up as a blue color in astrophotography.
Type II
This is usually the brightest tail visible. It is composed of larger dust particles streaming from the comet. If the comet is moving fairly rapidly in its orbit, this tail is likely to be curved. The color of the tail is usually white, but can also appear slightly grey or yellow from reflection of sunlight.

Look at the tail(s) in more detail. Is there one, two or even more tails visible? Can you see any structure in the tail(s)? You can quite frequently see knots within the tail and the solar buffeting quite often produces disconnection events, where parts of the tail become detached from the comet. Can you see a difference between the ion and the dust tail? Are the tails pointing in different directions? How does the view differ at different magnifications or different instruments? For large, bright comets try observing it with wide angle binoculars or a rich-field telescope to bring out subtle detail around the comet or in the tails that you cannot see in close-up telescopic views? If it is bright enough to be seen with the naked eye comet, how does it look when observed that way?

The developing tail is forced out from the comet by the radiation coming out from the Sun. Therefore any tails that the comet has will almost always point in the opposite direction to the Sun. The rapid movement of the comet along its orbit and the geometry of our view may distort this somewhat, producing tails that are curved, or even multiple tails. Some of these tails can even be observed pointing in opposite directions and even towards the Sun. This is solely down to the separation of the ion and dust tails and the angle of our particular vantage point at the time.

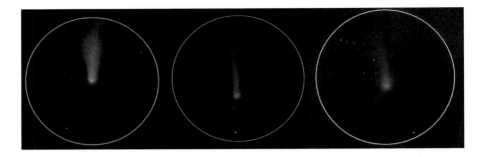

Fig. 11.4 Drawings of comet PANSTARRS C/2011 L4 showing its changing appearance in March, April and May 2013 (Courtesy of Seb Jay)

Background Objects

Sometimes a comet will pass background objects as it moves across the sky.

In February 2012 Comet Garradd C/2009 P1 passed close the bright globular cluster M92. Comet PANSTARRS C/2011 L4 passed close to M31 in March 2013. Both of these events gave opportunities for viewing or capturing great images.

What about stars or other deep-sky objects that might be shining through the comets head or tail? Are they distinct?

11.7   Predicting a Comets Future Appearance

Many of the graphical planetarium software packages used to predict the appearance of a future comet do attempt to show the appearance of the comet in the sky. Nevertheless, this can only ever be a representation of the comet, and will never really show how it will actually look. The tail will always be shown pointing directly in the opposite direction to the Sun (anti-solar tail). This might be so for the much lighter ion tail, but the dust tail, especially if the comet is close to the Sun and moving fairly fast along its orbit, could be markedly curved. Any predicted appearances of comets can be way off the mark when the comet does eventually appear in our skies.

We cannot predict with any real certainty how bright a particular comets tail will be, whether it will have a tail (or multiple tails) at all until it start to perform. Really looking at comets and taking in the view will reward the observer with seeing wonderful structure within. Comet Hale-Bopp was a fine sight through a large telescope. Its head showed a radial series of dust that had been ejected from the nucleus that could be seen rotating over a few hours. So it's always highly rewarding to look at these objects with different instruments and different magnification to see everything that the object has to offer. In addition, take your time. The comet is being constantly buffeted by the Sun's radiation, so it is continuously changing.

In 1973 Comet Kohoutek C/1973E1 was discovered well beyond Jupiter and was considered to become extremely bright later that year. It was indeed hyped in the media as "The Comet Of The Century". Needless to say it didn't quite live up to expectations. It still achieved naked eye brightness, but to the public it was certainly disappointing. Also disappointing to many was the return of Halley's Comet in 1985/86. Comet Hale-Bopp made up for both of these in 1987 when it was easily visible to the naked eye for many months. At the time of writing we are awaiting the arrival of Comet ISON C/2012 S1 at the end of 2013, where it has now been dubbed "The Comet of a Lifetime". Recent observations show it to be underperforming some months before its arrival. At the time of publishing ISON has recently been recovered from behind the Sun and is still looking to give a reasonable display in late 2013.

As David Levy once commented:
Comets are like cats: They have tails and they do precisely what they want.

Comets can therefore be notoriously tricky to predict with certainty. It really is no wonder many seasoned comet observers tend to play down the future predictions of a comet's performance. That is one of the things so fascinating and tantalizing about these fascinating objects. General notes about the possible future appearance of a comet to be aware of are listed below.

The closer a comet approaches the Sun, the bigger and brighter its coma becomes more active and more material is ejected. The tail(s) will also become much longer.

A "new" comet from the speculated Oort cloud (If theory is correct) have never experienced solar heating before. These have the potential of becoming very active and much bigger and brighter as they approach the Sun.

The further a comet is discovered from the Sun, the greater the chance of it becoming spectacular as it gets nearer the Sun.

The larger the comet, the bigger and brighter it could become.

The nearer the comet comes to Earth, the bigger and brighter it could become.

A sun-grazing comet will appear at its brightest close to the Sun usually in the daylight sky.

11.8   Daylight Comets

A great comet, such as that in 1910, visible in the daylight is a very rare beast. Nevertheless, when it does happen it causes a lot of excitement, not only in the astronomical community, but also amongst the general public. Comet Hale-Bopp's appearance in 1987 made up for some disappointing comets in the past, but it still wasn't visible to the naked eye in the daytime, despite being a circumpolar object.

What brightness does a comet need to be to be visible in the daytime? John Bortle has put together a guide to judging the potential of what an observer might see in the daytime with comets of various brightness.

Magnitude −1.5

The comet can be located up to within 5° of the Sun with an 80 mm telescope if its precise position is known. Just visible. Not "obvious".

Magnitude −2.0

Visible with large mounted binoculars and easy in moderate telescope.

Magnitude −2.5

Seen with ordinary binoculars with care. Easy with modest telescopes. Faint hints of the tail may also be visible.

Magnitude −3.0

Amazingly bright in ordinary binoculars. At sunset it will be visible with the naked eye and very obvious in binoculars. Modest telescopes show possible structure in coma with Sun well above the horizon.

Magnitude −4.0

Just visible with naked eye if more than 5° from Sun. Telescopes show extremely bright coma. Visible in telescopes almost to Sun's limb (Caution!).

Magnitude −5.5 or brighter

Easily seen with the naked eye throughout the day if Sun blocked. A short tail may be visible.

Magnitude −7.5 and brighter

Immediately obvious on looking towards the Sun so may even be spotted by general public. A tail more than 1° in length may be visible. Significant internal structure within coma seen with a telescope.

John's Full PDF document can be downloaded from here:

http://www.eagleseye.me.uk/DaylightComets.pdf

11.9   Hunting for New Comets

This activity is a very competitive area. Anyone who discovers a comet and reports their observation before its discovery is officially announced gets the comet named after them. So what is the best way to go about to finding your very own comet? To join Charles Messier's favorite past-time you need dark clear skies, a low western or eastern horizon and plenty of patience. You also need to know the sky very well.

It may sound obvious, but you will also need to know what a comet should look like before you can identify them. How can you distinguish between a faint comet and a faint nebula/galaxy? This can only be obtained through practice by observing a number of comets.

On the other hand it is also beneficial to know what a comet shouldn't look like as well, so observing many different examples of deep sky objects will also allow you to differentiate. Like a lot of observing, pick your moments for comet hunting carefully. Large professional telescopes are constantly imaging the dark skies around midnight. Despite these NEO searches there is still an area of sky not covered by them that gives the amateur a greater opportunity to discover new comets. This is an area of about 80° around the Sun. Use this neglected area of sky for your hunting ground in your search for new comets to increase your chance of success.

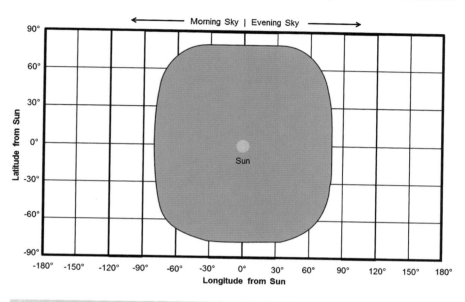

Fig. 11.5 Area of sky (*shaded*) not covered by current NEO searches (Adapted from: Can Comet Hunters Survive? Taki, T. & Murakami, S.)

Choose a time just after sunset, or before sunrise, when the Moon is well out of the way and the sky is still, dark and clear. Use a wide angle short focal length rich-field telescope.

To sweep for comets in the evening sky, about an hour after sunset start looking low the horizon towards the northwest This area of sky will soon set, so try and search this area first just above the north western horizon. Sweep using a smooth motion southwards parallel to the horizon and do not hold your breath! This will reduce the amount of oxygen reaching your brain and the light sensitivity of your eyes will be reduced, so you won't be able to see such faint objects. Keep looking for anything that looks like it might be a slightly out of focus "star". When you get to the south west, move the telescope up about a third of the field of view and sweep back in the opposite direction until you reach just above where you started. Keep sweeping slowly backwards and forwards, about 80° from where the Sun set, then move up a third of a field up each time, sweeping backwards and forwards, looking for suspicious objects until you reach about 45° in altitude.

When sweeping for comets in the morning sky it is best to start higher up at about 45° altitude about 80° in the north eastern sky (South eastern in the southern hemisphere) a few hours before sunrise. Sweep in a similar fashion as in the evening sky, parallel to the horizon,. After each sweep move the telescope down a third of a field of view. Keep sweeping backwards and forwards until you reach the horizon. You can then sweep backwards and forwards as the sky rises towards you until twilight starts to interfere. It takes much patience as it can take hundreds of hours of observing to find a comet and not happen for many years. The rewards really are worthwhile if you have clear dark skies and lots of patience.

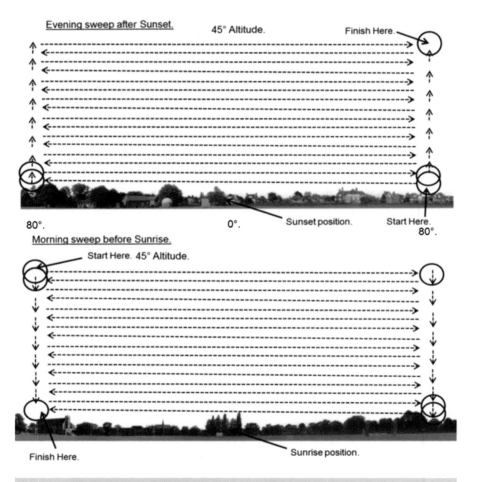

Fig. 11.6 Suggested areas for sweeping for new comets in the evening and morning sky (Courtesy of the author)

So what do you do if you think you have found a comet? Check the Internet, Star charts or planetarium software, making sure the comet details are up to date. Look in that area of sky to see if there are any deep sky objects or known comets currently in that part of the sky. Is it a really faint comet that has just had a sudden surge in brightness? This would be a very useful observation in itself. Can you identify your object with anything already known? If not, then you now need to delve a bit deeper. Can you follow it for an hour or two? Does it move over time? If it does, estimate its brightness and note the direction in which it is moving. Make a note of its position in the sky. Do you know someone who will take a look at the same part of the sky and verify your "discovery"? This can either be a friend or a member of one of the observational sections in astronomical organizations listed in the book. If after eliminating everything else, and have confirmation from another observer to back you up, you should report your observation. You can make reports to organizations like the BAA or directly to the International Astronomical Union Bureau for Astronomical Telegrams. If you are one of the first three people to report the find before the discovery is announced, the comet will be named after you. But beware! Sending in details about your new "discovery" that hasn't been confirmed and turns out to be anything but, isn't going to do your reputation and observing credentials any good whatsoever.

11.10   Comet Resources

Comet Hunting Primer.
  http://www.nightskyhunter.com/Comet%20Hunting%20-%20A%20Primer.html
  SkyHound Comet Observation Web Page.
  http://cometchasing.skyhound.com/
  Comet for Windows.
  Software to analyze light curves of comets to calculate possible brightness.
  http://www.aerith.net/project/comet.html
  Minor Planet Center Observable Comets page.
  http://www.minorplanetcenter.org/iau/Ephemerides/Comets/index.html
  Minor Planet Ephemeris Generator. Calculates the path and brightness of asteroids and comets as seen from a specified location.
  http://www.minorplanetcenter.net/iau/MPEph/MPEph.html
  Jet Propulsions Laboratory Horizons Web Page for calculating positions.
  http://ssd.jpl.nasa.gov/horizons.cgi#top
  How to report astronomical discoveries: IAU Bureau for Astronomical Telegrams.
  http://www.cbat.eps.harvard.edu/HowToReportDiscovery.html
  On occasions comets can pass close to other interesting celestial objects.
  A Comet rendezvous list is maintained by Seiichi Yosida.
  http://www.aerith.net/comet/rendezvous/current.html

Chapter 12

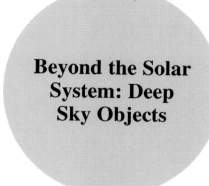

Beyond the Solar System: Deep Sky Objects

The observation of deep sky objects seems to separate some observers. Deep Sky enthusiasts enjoy hunting down faint fuzzy blobs, while others cannot see what all the fuss is about, instead relishing the planets and Moon.

There are a wide variety of deep sky objects, from the bright glowing nebulae, to faint distant galaxies. Planetary nebulae grace many constellations with their presence and fabulous globular clusters swarm around our own Milky Way galaxy as well as other more remote galaxies.

12.1 Observing Deep Sky Objects

Observing deep sky objects usually requires dark skies and large telescopes. Many of the brighter objects can still be seen reasonably clearly even if you have a light-polluted backyard, or you are in possession of a small telescope. There are some simple tricks that can be used to help you view faint deep-sky objects:

Averted Vision

This "trick" is used to get around the fact that the retina of the eye is less light sensitive in the center of our field of view. Averted vision relies on the higher light sensitivity of the rod cells around our peripheral vision to pick up the faint light of a distant object. To use averted vision you first find the correct field of view of the object in question, placing the object directly in the center of the field.

Look through the eyepiece. Concentrate on the area in which the object should be, but look towards a star towards the edge of the field of view. Quite often the

D. Eagle, *From Casual Stargazer to Amateur Astronomer: How to Advance to the Next Level*, The Patrick Moore Practical Astronomy Series, DOI 10.1007/978-1-4614-8766-1_12, © Springer Science+Business Media New York 2014

object, which was not visible before, often pops into view. Other objects, which can be seen directly can also look differently when using averted vision as fainter parts of the object spring into view.

The Averted Vision Scale

Ron Morales of the Sonoran Desert Observatory has written an Averted Vision Scale as shown below:

AV1 – Object can be seen with averted vision but once found, the object can occasionally be seen with direct vision. If an object is first noticed with averted vision but once found this object can then be seen steadily with direct vision it is considered a direct vision object as opposed to an averted vision object.

AV2 – Object can be seen only with averted vision but it is held steady. Here the sweep of one's vision makes the object detectable.

AV3 – Object can only occasionally be seen with averted vision as it "comes & goes" with the seeing conditions. In this case the object is seen more than 50 % of the time.

AV4 – Object can only occasionally be seen with averted vision as it "comes & goes" with the seeing conditions. In this case the object is seen less than 50 % of the time.

AV5 – Object can only be glimpsed with averted vision after a continuously viewing the field for a few minutes or more. This level of averted vision usually occurs when one carefully observes a field for a lengthy period of time. This might occur within the first 3–5 min of viewing the field. In this level it is important that the observer has no knowledge of the exact location of a possible object. Having such knowledge prior to viewing could mislead some observers into believing that they saw something they did not actually see. One problem associated with viewing extremely faint galaxies it that sometimes an extremely faint star could be misidentified as an extremely faint galaxy. For this level of averted vision it is suggested that the observer make a field sketch showing faint stars as well as the object in question. This field drawing can then, at a later time, be compared to an actual photograph or chart. At this level of detection are you seeing or just detecting the presence of an object?

Ron's Averted Vision Scale can be downloaded from here:
http://www.astronomylogs.com/pages/resource/pdf/13%20Averted%20Vision%20Scale.pdf

Other Deep Sky Observing "Tricks"

As well as using averted vision there are also a number of other methods observers have found work to help them see as much as possible.

Increase the Magnification

Putting in a lower focal length eyepiece sometimes gives you better contrast between the object and background sky. The eye is also much better at detecting larger, faint objects than small, faint objects.

Tapping the Telescope

When trying to observe very faint objects tapping the telescope often makes a faint object that is just on the edge of your visual perception pop into view. The slight wobbling movement helps your eye to see the faint light. Used in combination with averted vision, it is a very powerful method of getting the most from your scope.

Moving the Telescope

Moving the telescope slowly so that the object being observed moves across the field of view also helps to lift it into view. Some people switch off their tracking so that the Earth's rotation moves the field of view slowly in right ascension. Does moving the telescope slowly in declination have the same or better effect? Give it a try. What works for one person might not work for another.

12.2   Deep Sky Catalogs

How do you start hunting down these objects? There are many catalogs available for the deep sky observer that will lead you to these remarkable objects and help identify them.

Familiar Names

Some of the most interesting deep sky objects have their own unique names. These include the favorites, The Ring Nebula (M57), The Whirlpool Galaxy (M51) and The Eskimo Nebula (NGC 2392). It is always useful to remember some of these common names. The vast majority of deep-sky objects are contained within a number of catalogs. There are now a bewildering number of catalogs available, of which many objects appear in more than one.

The Messier Catalog

This is probably the most famous catalog for the deep sky observer, and one most observers starting out in the hobby are familiar with. Charles Messier was mainly interested in hunting for comets. He frequently came across some fuzzy objects that did not move that could be mistaken for comets. He started to compile a list of these "nuisance" objects, so they could be avoided. His legacy to the observer is his list of bright deep sky objects. The list has been added to by other observers over the years resulting in 110 objects for the amateur to enjoy.

An online Messier Catalog is maintained by the Students For The Exploration And Development Of Space (SEDS).

http://messier.seds.org/?gclid=CJ3DzvPXlLYCFXDKtAodF3QAJg

The Messier Marathon

As the Messier catalog only contains 110 objects it is a small enough selection of objects for all to be observed in one night. Of course conditions have to be just right to be able view all the objects in the catalog during that single session. The key to success, like most observing is in thorough planning. Make a list of the objects you will observe and in which order. Choose wisely. The best time of year to attempt a

Messier marathon is mid-late March or early April. March being the optimal month for an attempt, but early April can also be used. You must also be located in mid-northern latitudes to be able to see all the objects as some of them are quite a way towards the south. The complete Messier marathon can really only be attempted between 10° and 35° N where all the objects rise above the horizon during the night. If you live above or below these latitudes you will have to make do with setting your target a little lower.

The marathon starts with a mad scramble as dusk falls to capture the brighter Messier objects now starting to set over in the west. Once the skies darken sufficiently, track down the other fainter Messier objects over in that direction. The rest of the nights observing is a little more relaxed as the patient observer ticks object after object off their list. The pace is always fairly high, especially as tiredness creeps in later during the night. In the morning sky, the Virgo area is so easy to get lost in amongst all the faint "fuzzies", so the observer has to move steadily and systematically through the sky. As dawn approaches it's another scramble to catch objects rising before dawn starts to break. Anyone who is still awake enough after an all-night session and manages to spy all 110 objects really deserves the utmost respect.

Most marathon participants set themselves a target of a number of objects just in case they fail to find some of them. You can always have another attempt another year to try and improve on your previous total.

Refer to this Web page for an explanation and guidance for conducting a successful Messier marathon:

http://messier.seds.org/xtra/marathon/marathon.html

If you are planning a Messier Marathon yourself, Larry McNish's Messier Marathon planner can be used to produce an object observing list order for the date you plan your marathon:

http://members.shaw.ca/rlmcnish/darksky/messierplanner.htm

There is also a list of Messier Marathon tips available on this page:

http://www.maa.clell.de/Messier/E/Xtra/Marathon/mm-tips.html

The Year-Round Messier Marathon Field Guide. Pennington, H. C. Willmann-Bell, 2008.

The Observing Guide to the Messier Marathon. Machholz, D. Cambridge University Press, 2002.

Atlas of The Messier Objects. Stoyen, R. Cambridge University Press. 2008.

The Caldwell Catalog

A much more recent catalogue that has appeared in the last few years is the Caldwell Catalog. This is a list of 109 fairly bright objects that are not included in Messier's catalog. Unlike Messier's list, this catalog contains objects right the way across the sky in both the northern and southern hemispheres. This list was compiled by the late British astronomer Patrick (Caldwell) Moore. Do we need another observing catalog? Whatever your views on the appearance of this list, it does bring many interesting and quite bright objects to the attention of the more casual observer that may otherwise get overlooked.

Caldwell Catalog List:
http://messier.seds.org/xtra/similar/caldwell.html
Astronomy League Caldwell Catalog Page:
http://www.astroleague.org/al/obsclubs/caldwell/cldwlist.html
The Caldwell Objects and How to Observe Them. Martin Mobberley. Springer 2009.

The Herschel Catalog

William Herschel compiled a list of over 2,500 deep sky objects. Thankfully we don't have to dive in and look for this many objects, unless you really want to, as his catalog usually appears as the Herschel 400. This lists just his brightest and most interesting objects.

Observing the Caldwell Objects. Ratledge, D. Springer. 2000
The Herschel Objects and How To Observe Them. Mullaney, J. Springer. 2007.
A Complete Guide To The Herschel Objects. Bratton, M. Cambridge University Press. 2011

The NGC Catalog

The NGC Catalog first appeared in 1880. Numerous revisions later the catalogue now contains 7,840 objects. These were originally numbered according to right ascension. So NGC 1 just on RA of 0 h, and NGC 7,840, close to 24 h right ascension (although due to precession their RA has now moved, so this is no longer the case).

An interactive NGC (Also IC and Messier) catalogue is maintained by SEDS.
http://spider.seds.org/ngc/ngc.html
Of course these catalogs, despite due diligence and the amount of care put into compiling them, the data held is not necessarily as accurate as we would like them to be. This is especially so if the data is from original observations made many years previously.

The NGC/IC Project is a collaborative effort of professional and amateur astronomers. This is attempting to correctly identify any misplaced or described objects and reduce the number of errors they contain:
http://www.ngcicproject.org/
There is, of course a large overlap in objects that appear on more than one list. There are also a number of other deep-sky catalogues that you will come across over time. Some of these are mentioned later in this chapter when we come to discuss the different types of objects. It is quite clear that there are many deep sky objects for the amateur astronomer to enjoy. Let's have a quick look at some of the types of deep sky objects that can be observed.

12.3 Open Star Clusters

There are some glorious star clusters in the night sky. These are mainly spread along the plane of our Milky Way. Therefore, the best times to observe these are in the Summer, or the dead of winter when the Milky Way is high in the sky. The stars that comprise the cluster are quite often formed from the same nebulae and share a common motion through space. The stars are usually fairly loosely packed with plenty of space between each star. They are usually fairly old clusters.

Classification of Open Star Clusters
Open star clusters are classified according to the method developed by R.J. Trumpler. Three indicators that are measured:
Concentration and detachment from the surrounding Star field.

Class I: The cluster is strongly detached from the stellar background with a strong core stellar density.
Class II: The cluster is detached from the stellar background with a light core stellar density.
Class III: The cluster is detached from the stellar background without a denser core.
Class IV: The cluster is weakly detached from the stellar background, the area having a higher stellar density but no visible core.

Range in brightness of the stars.

Class 1: All the stars present about the same brightness.
Class 2: The stars present a regular range of brightness.
Class 3: Beside some very bright stars, many weaker stars with a wide magnitude range.

The number of stars the cluster contains.

p = The cluster is poor in stars (less than 50 stars)
m = The cluster has a medium number of stars (from 50 to 100 stars)
r = The cluster is rich in stars (more than 100)
n = A nebula is linked to the cluster.

So a couple of typical open star cluster classifications would be:

M11, The Wild Duck Cluster is scored: I 2 r.
NGC 2818 is scored: III 1m n.

12.4   Globular Star Clusters

These spectacular gatherings of stars are really densely packed. The observer will be hard pushed to see any space between its component stars, especially towards the middle if the cluster. These appear to surround the Milky (and indeed other galaxies) and widely vary in their distance to us. The grandest globular cluster in our sky is undoubtedly Omega Centauri. It is located in the southern hemisphere, so is not observable from more northern latitudes. In the northern hemisphere the nearest there is to this behemoth is the Great Hercules Globular Cluster, M13.

Globular clusters are classified according to the degree in which the stars within the cluster are concentrated using The Shapley-Sawyer Classification. There are 12 classifications within the scale, each one using Roman numerals. (The visible differences between each neighboring classification is sometime very subtle).

I – III = Very high stellar density at the core. Halo decreases in luminosity as a function of the distance from the core.

IV – VI = Stellar core still visible, but more spread out and less dense.

VII – IX = Stellar density is more homogenous and less contrasted.

X – XII = Cluster luminosity completely homogenous and no increase in stellar density at the core.

So take time to hunt down some of these fabulously densely packed collections of stars. Something to look out for while observing globular clusters is to look for chains of stars. M13 has a number of these making it somewhat reminiscent of spider's legs. There may also be some vague shapes visible within the clusters structure where the concentration of stars is higher or lower. The most famous of these is the slightly darker "propeller" shape that is also visible within M13. Some observers seem to be able to see this structure easily, others less so. Does it appear for you? Does aperture or magnification help you to view it?

Globular clusters not only surround The Milky Way, they also swarm around other galaxies. It should therefore come as no surprise that we can also observe them around a few of our nearest and brightest neighboring galaxies. The Andromeda Galaxy M31 has a number above 15th magnitude, which can be observed directly if you have a large telescope. The brightest at magnitude is known as G1 (Mayall II) and is located about 2.5° to the south west of the nucleus of M31. At magnitude 13.8 it will require a fairly large telescope above 10″ to find where it forms a triangle with two stars of 14.5 and 15th magnitude.

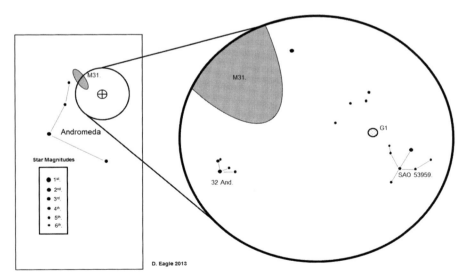

Fig. 12.1 Finder chart for the globular cluster G1 around M31, the Andromeda Galaxy (Courtesy of the author)

Some observing notes and a more detailed map for finding G1 are located on this Web site:

http://deepsky.astroinfo.org/And/g1/index.en.php

Exercise 12A: Picking Out Globular Clusters Around the Andromeda Galaxy

As well as G1 there are also a number of globular clusters bright enough to be found around M31. These really do need 10″ or bigger telescope. These globular clusters are a bit more challenging than G1 as they are situated in front of the galaxy itself. This results in the background sky being a little brighter, so that there is less contrast.

A monochrome finder chart with the globular clusters marked around M31 is available from the Cloudy Nights Web site:

http://www.cloudynights.com/images/SWandromedaglobs/m31globs_inverted _bw.jpg

12.5   Planetary Nebulae

These nebulae are formed from extremely powerful stars that are getting towards the end of their life. In their struggle to survive they have thrown off their outer atmosphere This produces a bubble of gas surrounding the star. Hubble space telescope images reveal beautiful patterns in these objects, show how prevailing conditions around the star that created them has crafted the gas into wonderful shapes. These nebulae are most abundant along the plane of the Milky Way. The area around Aquila is especially rich with them, making a great hunting ground for these faint puffs of gas from July until late in October. Planetary nebulae emit a lot of radiation from glowing ionized oxygen. If you have an OIII filter, you can alternatively pass the filter between the eyepiece and your eye. Any stars in the field of view will fade when the filter is in place, but any planetary nebula will stay about the same brightness. Unlike the stars, most of its light passes unimpeded through the filter. A simple filter makes identification of even very small planetary nebulae relatively easy.

The planetary nebula that we normally observe in the telescope is only the brightest part of the gas shell. There is usually a more tenuous and much fainter outer shell of gas surrounding the nebula. This is extremely difficult to observe, mainly shining in the Hydrogen Alpha. But they do record on long exposure astrophotography using hydrogen alpha filters.

Planetary nebulae are classified according to their structural characteristics on something known as the Vorontsov-Velyaminov Scale:

1	Stellar image. (Looks similar to a pinpoint star)
2	Smooth disk
2a	Smooth disk. (Brighter toward center)

(continued)

(continued)

2b	Smooth disk. (Displaying uniform brightness)
2c	Smooth disk. (Shows traces of a ring structure)
3	Irregular disk
3a	Irregular disk. (Very irregular brightness distribution)
3b	Irregular disk. (Traces of a ring structure)
4	Ring structure
5	Irregular form. (Similar to a diffuse nebula)
6	Anomalous form. (No regular structure)

12.6   Gaseous Nebulae

These nebulae are mainly divided into three different types: Emission, Reflection and Dark. All nebulae are by and large exactly the same thing, composed mainly of hydrogen gas and dust. These are areas where stars are being created in a huge cosmic nursery. The only thing that changes is the surrounding conditions in which the nebula happens to find itself. The clouds of gas are essentially dark, unless something like a close-by star, or a star within it that erupts into life, happens to supply them with enough energy to ionize the gas.

Emission Nebulae
These are nebulae that exist in star-rich areas of sky where stars have already been created. The gas has been strongly ionized by the radiation of the stars within them. This causes the gas to shine with a distinctive red light. Unfortunately for us, our eyes (and standard DSLR's) are not very sensitive at this end of the spectrum. As a result most emission nebulae appear fairly faint in the telescope and take a fairly long time to record on images. There are of course a number of exceptions to the rule, the wonderful Great Orion Nebula (M42) being the most famous example of this type of object as it is so very bright.

Reflection Nebulae
These are formed by stars lying above the plane of the gas cloud, closer to the Earth. In this case the gas isn't ionized by the star. The light from the foreground star is reflected back off the gas towards us. This creates a very faint glow, which has a distinct blue color on long exposure photographs. This is in stark contrast to the brighter emission nebulae. An example of this type of nebula is M78 in Orion.

Dark Nebulae
These are clouds of gas and dust that are not ionized by nearby stars, nor are there stars nearby that will cause light to be reflected off them. They are usually revealed by the fact that they block off the light from objects behind them in the background. There are many dark nebulae catalogued. E. E. Barnard in particular did a lot of work in this field in the early 1900s. There are many examples of this type of nebulae in the Milky Way, where they are easily identified when they block off the faint

haze of background Milky Way stars beyond. These are notoriously difficult to observe, as we are relying on a bright background object that will help to show them in silhouette. The most famous example of this type of nebulosity is the Horsehead Nebula. The Horsehead is quite a difficult object to observe, so let's try and observe another object of this type in Aquila that is a bit easier to spot and is visible from most of the Earth.

Exercise 12B: Observe the Dark E in Aquila

There is a cloud of dark dust lying close to the bright star Altair in Aquila. It is visible due to blocking out the light of the more distant background stars from the Milky Way. This object is best observed in dark skies. This particular dark cloud is No. 142 and 143 in Barnard's catalogue of dark nebulae. This has been said to have been spotted with the naked eye, but excellent clarity, dark skies and excellent night vision is required. This dark nebula covers an area 2×1 degrees of arc, so is fairly large. It is therefore best observed using a rich-field telescope or a pair of binoculars.

E.E. Barnard's list of dark nebulae is freely available on the Web having been reproduced by the Saguaro Astronomy Club.–The Best of Barnard's Dark Nebulae.

http://www.saguaroastro.org/content/Best-of-Barnards-Dark-Nebulae.htm

His original two-volume publication has recently been re-published as a single hard-back edition:

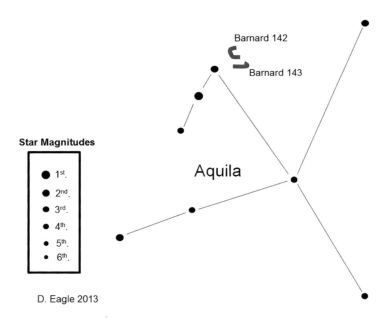

Fig. 12.2 Finder chart showing the location of the dark nebula, Barnard's E in aquila (Courtesy of the author)

A Photographic Atlas of Selected Regions of the Milky Way. Barnard E.E. Cambridge University Press. 2011. Originally published 1927.

12.7   Supernova Remnants

There is only one supernova remnant listed in Messier's catalogue, this is the first entry in his catalogue, M1, The Crab Nebula. After its discovery it was soon linked to a supernova observed in 1054 in Taurus by Chinese astronomers.

These are the remains of stars that have reached the end of their lifetime and have literally torn themselves to pieces, in a supernova event, spreading material out into surrounding space. Thanks to supernovae like this, the heavier elements produced in the hearts of developing stars have been seeded into space. This has resulted in later generations of planetary systems like our own being liberally sprinkled with just the right amount of elements required for life to exist. So we do really owe our very existence to stars from many millions of years ago that have long faded from view.

The younger the supernova remnant is, the more concentrated its surrounding cloud of debris is. Subsequently it follows that the smaller it is, the more compact the object is, therefore and the brighter and more easily observed it will be. The older the supernova remnant, the more spread out it is, the fainter the object. The Crab Nebula being a relatively new remnant is still quite a compact object. The Veil nebula being much older, the star that formed it exploded between 5,000 and 8,000 years ago. It is still bright enough to be viewed through any telescope above 10 in., but is much more spread out and harder to observe. There is another supernova remnant located on the border of Taurus and Auriga called Simeis 147. The supernova that caused these gradually spreading arcs of light went supernova approximately

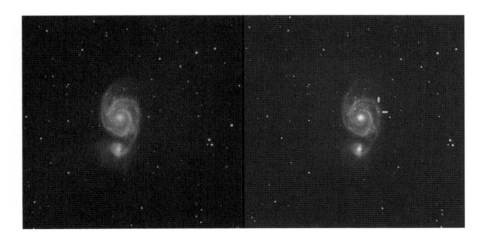

Fig. 12.3 Supernova 2011 DH located in a spiral arm of M51 (Courtesy of the author)

40,000 years ago. The faint filaments that comprise the object today are extremely faint and is very difficult to observe. They have however been captured using astro-photography in dark skies.

There are a number of other supernova remnants that have been discovered listed in Wikipedia:

http://en.wikipedia.org/wiki/List_of_supernova_remnants

12.8   Galaxies

There are many external galaxies visible to the amateur astronomer. The brightest are the Large and Small Magellanic Clouds, (after our own Milky Way, of course). These two satellite galaxies are only visible from the southern hemisphere. The next brightest is the Great Andromeda Galaxy M31. This is a naked eye object, given favorable skies, although The Triangulum Galaxy M33 isn't that far behind in brightness. All of these are extremely close to us (on the cosmic scale). With careful observing and clear dark skies, the deep-sky enthusiast can enjoy the view of faint galaxies many millions of light years away. The very thought that a photon of light leaving a galaxy well before the age of the dinosaurs, or even before the Earth's crust cooled enough to form a solid crust, has travelled all that way, finally ending up hitting the back of your retina with just enough energy to cause a small chemical reaction just so you detect it as a faint fuzzy patch. Miraculous!

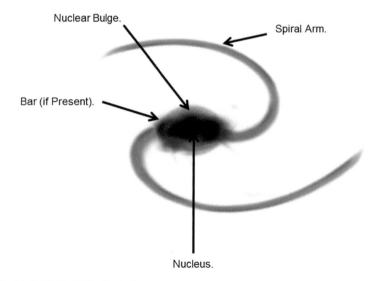

Fig. 12.4 Features typically found in a galaxy (Courtesy of the author)

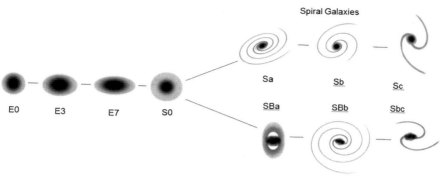

Fig. 12.5 Edwin Hubble's tuning-fork diagram (Courtesy of the author)

Edwin Hubble was the first person to discover that galaxies were distant "island universes" and started to systematically classify galaxies. He devised the "Tuning Fork" diagram, which was meant to show some sort of evolution of galaxies, moving from left to right as they evolved. Immature elliptical galaxies on the left, older spiral galaxies on the right. We now know that this is incorrect and although the diagram is outdated, it is still widely used to classify the shapes of the galaxies we can observe.

The different types of galaxies and the nomenclature he used in his diagram are still used today. In fact astronomers today still refer to elliptical galaxies as "early Galaxies" and spiral galaxies as "late galaxies".

E = Elliptical Galaxies

These are categorized in eight classes, ranging from E0 right the way up to E7 depending on how elliptical the galaxy is shaped.

E0 = Round shaped galaxies.
E7 = Extremely elliptical (Virtually flat).
S0 = Lenticular Galaxies

These galaxies are in a class of their own.
They display a central bulge and disk, but have no visible spiral arms.

S = Spiral Galaxies

These galaxies have spiral arms that are attached directly onto the central nucleus.

SB = Spiral Barred Galaxies

These galaxies have spiral arms that are attached to a central bar.

These last two types of spiral galaxies are also divided into three groups, depending on how tightly the spiral arms are wound:

a = Arms wound tightly with large central bulges.

c = Arms loosely wound and have small central bulges.

b = Somewhere in between the last two classes.

De Vaucoulers System of Galaxy classification has later been added to Hubble's original system. This includes a number of other values.

Bars

The presence or absence of a nuclear bar. Although SB galaxies would represent a bar being visible, there are cases where there are intermediate types for weakly barred spirals (SAB). Lenticular galaxies can also be classified as unbarred (SA0) or barred SB0). S0 is used for galaxies where the presence of a bar cannot be determined due to the presentation towards Earth.

Rings

r – The galaxy shows the presence of ring-like structures.

s – The galaxy shows no ring like structure.

rs – The galaxy exhibits some transition between the two.

Spiral Arms

Sd (SBd) – Spiral arms diffuse and broken, made up of individual stellar clusters and nebulae.

Sm (SBm) – Spiral arms irregular in appearance. No bulge component.

Im – Highly irregular galaxy.

Using this classification, some galaxy types are listed as:

M51 – Sab(rs)cd.

M31 – SA(s)b.

M82 – I0.

M64 – (R)SA(rs)ab.

The Yerkes Scheme

Galaxies are also classified according to their star spectra results of the inner part of the galaxy, their morphology and their presentation towards Earth.

Spectral Type

a – prominent A stars.

af – prominent A-F stars.

f – Prominent F stars.

fg – Prominent F-G stars.

g – Prominent G stars.

gk – Prominent G-K stars.

k – Prominent Kstars.

Galactic Shape

B – Barred Spiral.

D – Rotational symmetry with no pronounced spiral or elliptical structure.

E – Elliptical.

Ep – Elliptical with dust absorption.
I – Irregular.
L – Low surface brightness.
N – Small bright nucleus.
Q – Quasi-stellar objects (unresolved high red-shift galaxies).
S – Spiral.

Inclination
  Graded from 1 to 7, where:

1 = Face-on to Earth.
7 = Edge-on to Earth.

The Andromeda Galaxy M31 would be classified using this system as kS5.
What do you see when you observe a galaxy? Look at the nucleus. Is it round, irregular or elongated. Is it evenly illuminated right the way across? How about arms, or knots within them?

Scattered around the sky are also collections of galaxies, which are gravitationally attached. The nearest galaxy cluster is our local group, containing us, M31 and a number of other smaller galaxies. The next nearest galaxy cluster is in Virgo. As you extend your travels out into the void, there are many other cluster galaxies to explore. Stephan's Quintet in Pegasus is always a neat object to find a little south of NGC 7331. But if you want take on something a bit more challenging try finding some of the remote Abell galaxy clusters.

12.9   Quasars

These are extremely distant galaxies that are only visible from this distance due to the fact that they generate an enormous amount of energy. These objects are very compact, some no bigger than our solar system.

The brightest quasar visible in the sky is 3C 273 in Virgo. At magnitude 13 it lies at a distance of two billion light years. If you have a telescope at least 10″ in diameter, see if you can find this remote searchlight.

This quasar lies close to a small but distinctive "W"-shaped pattern of stars, looking much like a mini Cassiopeia. Download detailed instructions and a map of its location from my Web site. www.eagleseye.me.uk/Guides/3C273.pdf

12.10   Deep-Sky Resources

Atlas of the Milky Way:
  http://ned.ipac.caltech.edu/level5/ANDROMEDA_Atlas/frames.html

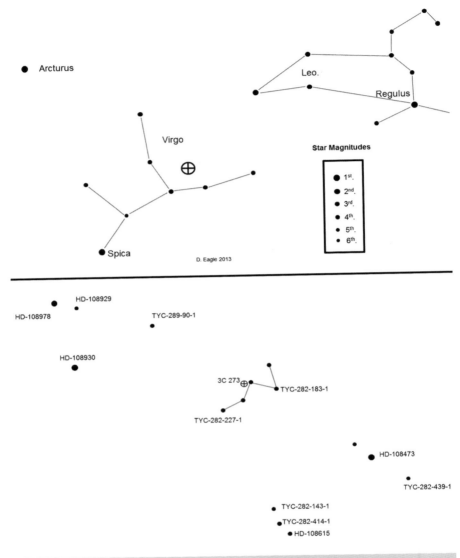

Fig. 12.6 Finder chart for quasar 3C 273 in virgo. *Top* wide angle view. *Bottom* close-up view (Courtesy of the author)

The Webb Deep-Sky Society. Dedicated to encourage the observation of Double Stars and Deep Sky Objects. Produce a useful collection of star maps and observing guides. http://www.webbdeepsky.com/

Astronomical Leagues Herschel 400 Web Page: http://www.astroleague.org/al/obsclubs/herschel/hers400.html

Abell Galaxy Clusters. A list of observable galaxy clusters from Abell's catalogue with downloadable images.

https://www.cfa.harvard.edu/~dfabricant/huchra/clusters/images.html

Arp Galaxies. A List of Halton Arp's 15 faint Peculiar Galaxies.

http://arpgalaxy.com/

Palomar Globular Clusters. A list of 15 faint globular clusters discovered on the Palomar Survey plates.

http://www.astronomy-mall.com/Adventures.In.Deep.Space/palglob.htm

Faint Fuzzy Deep Sky Observing Guides.

http://faintfuzzies.com/DownloadableObservingGuides2.html

The Night Sky Observers Guide: (2 Volumes). Kepple, G & Sanner G. Willmann Bell. 1998.

Deep Sky Catalog. An online List of the Index Catalogue(IC), New General Catalogue (NGC) and Uppsala General Catalogue of Galaxies (UGC) Objects.

http://www.skyfactory.org/deepskycatalogue/

Chapter 13

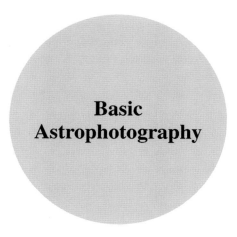

Basic Astrophotography

Taking images of the night sky can be extremely rewarding, producing beautiful images that provide a permanent record of what you have seen. Often known as "the dark side of astronomy" to the beginner, even to some seasoned astronomers, astrophotography does appear to be very much a dark art. Mention of stacking images, flat-fielding and light frames seems to talk of a highly complicated process and can send many people scurrying off for cover long before they have even started.

But when you are starting out, it doesn't need to be very complicated for you to produce your own rewarding images. Really pleasing results can be achieved with nothing more than a camera on a tripod and the minimum of processing.

13.1 Astro-imaging Background

A DSLR is the preferred camera of choice, but if your budget is tight and you already have a pocket camera, this can be used as long as it can do long exposures of between 10 and 30 s or longer. If the shutter is opened for only a short time, and/or use a higher focal ratio setting is used, then the image will be darker. Images taken like this could possibly be under-exposed. If the shutter is opened for a long time and/or a lower focal ratio setting is used, then the image will be brighter. Images taken like this could possibly become over exposed.

A very long exposure of a bright object will definitely over-expose the image. Too many pixels become white and no contrast is possible between different bright areas as they all look white. Too short an exposure and everything is very dark with no contrast and you will see lots of noise. To take successful images you need to get the exposure just right to give a good brightness range across the image.

Most objects in the sky are extremely faint. As they send so little light our way this means, that to reveal them in all their glory, we will need to take long exposures to build up the amount of light that is collected. The longer the exposure you use, the brighter the object becomes and the fainter the objects that reveal themselves on the image. The more light you collect the better the quality of the image you will eventually end up with. To achieve this you will need to keep the camera's shutter open for a long time.

Added to this, all cameras produce a certain amount of noise. By taking a series of images exactly the same, we can use software to add all the images together. This is called stacking. This smoothes out the noise and produces a much cleaner image. Some well-seasoned astrophotographers take a huge number of long exposures and stack them together in this way. This means that data can be collected over a number of nights, or even over a few years to collect data of a very faint object that is good enough to produce a very high quality image.

13.2 Basic Astrophotography Primer

Taking images of the night sky doesn't have to be complicated and you don't need any specialist equipment to do so. Following this guide will allow you to produce images for yourself without having to take out a second mortgage just to take many fabulous and rewarding images. It can be done using the simplest of photographic equipment. So what's needed?

1. A Camera

Any camera that is capable of taking an exposure of more than a few seconds exposure is capable of taking images of the night sky.

A DSLR has far more flexibility in what you can achieve, so is the preferred option if you can afford one. There is a lot of dispute on the Internet regarding the different makes of DSLR that are ideal for taking astronomical images. Most people seem to regard the Canon range as being the best, with lots of talk of the Nikon stripping out data when it takes an image. If you have a Nikon or other make of camera, excellent images can still be taken regardless. Many fabulous images have been taken with Nikon cameras, so there is no necessity to change camera if you already have one to hand.

There is a lot of third party software available for the Canon range, so if you are considering buying your first camera, and want to use a lot of different software to run your camera, a Canon is recommended for this reason only.

Make sure that you set your camera to not use long exposure noise reduction. Switch it off in the camera's menu. Also, save the resulting image as a RAW file if possible. (You may also want to save a JPG at the same time as a RAW, you will see why later).

Fig. 13.1 The authors trusty DSLR and shutter release timer (intervalometer) for taking multiple images (Courtesy of the author)

Set the ISO setting to at least 400 or preferably more (800–1,000). The higher the ISO setting the brighter the resulting image and the fainter the object you record in less time, but you will get more noise on your image. The lower the ISO setting the darker your image will be and the longer the exposure needed to see faint objects, but you do get a bit less noise on your image. As you can see it's a real juggling act getting it just right.

2. Lenses

The lens you use will determine what you capture. A wide angle lens ~20 mm will have a wide field of view and show a wide expanse of sky. A 100 mm lens will show a much smaller field of view and a smaller portion of sky. The shorter the focal length of lens used, the brighter the image and the less star trailing is visible. The longer the focal length, the fainter the image and the more star trailing will be visible. A bigger F-number will produce a fainter image, and will need a longer exposure. A smaller F-number will produce a brighter image, but will show up any light pollution very quickly on the image. A variable focal length lens will enable you to make framing of different constellations much easier but the quality of the lens and the f-ratio is normally less, so could affect the quality of the resulting image.

3. A Sturdy Tripod

Holding the camera steady is essential. You need something that will hold the camera steady for the extent of the exposure. For ease of use a full sized tripod is ideal.

You don't necessarily need a full sized tripod in order to achieve this. A small table-top tripod will suffice or a Gorilla Pod, or something like it, can also be used. A mount like this is light and portable, so is ideal for travelling. It will enable you

Fig. 13.2 A static camera mounted on a tripod for un-driven imaging (Courtesy of the author)

to mount the camera on any convenient sturdy support. Many a travelling astrophotographer has managed to take un-driven Milky Way shots with a DSLR on holiday this way, with the camera supported on the back of a Sun lounger or deck chair.

If the lens you are using has image stabilization built in and you are using your camera on a tripod, make sure that you switch the image stabilization off. If you leave it on it will produce vibrations of its own and ruin your images.

4. Multiple Exposure Capacity

Many cameras are only able to make exposures of up to 30 s on their own. To take images of faint objects, you need to make exposures of a minute or more. To achieve this, you can buy accessories that can be used to make longer exposures. These can be as simple as an infra-red release, which will open the shutter when pressed, and close the shutter when subsequently pressed. An attachment for the camera can be picked up for a reasonable cost these days, especially if you avoid the camera branded models. They can be set to get the camera to take a number of images one after another. This method of taking images is almost essential unless you want to stand there and keep releasing the shutter manually. Many an astrophotographer knows from experience, that this does become somewhat tedious after

Fig. 13.3 Using a gorilla pod as stable imaging support on a sun lounger when traveling (Courtesy of the author)

a few hours. So at least once past the beginning stage a way of taking images automatically must be used.

13.3 Taking Your First Images

Now we have really come onto the business end of things and are ready to take our images. Let's assume you've got your camera mounted on a tripod and are out under a clear sky. Pick an area of sky you are interested in and point your camera towards a familiar constellation. If you have a variable focal length lens frame your chosen area of sky accordingly.

Looking through the viewfinder, at this point many people are surprised to find that they will quite often only be able to see the brighter objects. So to see what is included in the shot you will need to take some test exposures. This will also help you to get the camera focussed.

Focusing the Camera

Take an exposure of about 10–30 s then look at your resulting image. Despite setting your lens at infinity you will more often than not find that the stars are more than pinpricks and will not be exactly in focus. You may need to experiment a few times and keep adjusting the focus to get really pin sharp stars. Once focused, tape the focusing ring so that it cannot be moved by accident. Once you have achieved proper focus, you can now start to take your images, but we are not yet finished with setting things up. Take a number of exposures of the same object and vary your exposure in each one. Take images from 10 s up to 1 min exposures, increasing each exposure by 10 s each time. Inspect each image, zooming in if necessary, or transfer the image to your computer, and note at which exposure star trailing starts to become apparent.

You will have then found the maximum exposure you can take before trailing is visible with that focal length lens. (As a rough guide if you use a 20 mm lens you can get away with about 45–50 s exposure before star trailing shows up as a result of the Earth's rotation starts to show). Now that the camera is set to take our images we can now start to capture the night sky. There are a couple of different approaches we can take.

Taking Single Images
This is by far the simplest form of astrophotography. Point the camera at the area of sky you are interested in imaging. Check the view through the camera and adjust the lens so that the image is framed correctly. If you have a zoom lens, try zooming in or out and see what effect that has. Try to include something in the foreground to give the scene some sort of scale and interest. Point the telescope at an area of sky you are interested in. Once focused, as before, take a single exposure of between 10 and 30 s.

Taking Multiple Images for Stacking
Using the same steps shown above frame the area of sky you want to image. Without changing anything between them take a number of images all exactly the same, keeping the camera pointed at the same chosen point in the sky. If you do change the focal length of the lens, or change lens you will need to check focus once again before continuing.

13.4 Processing Your Hard Earned Images

After a few hours you should have a few sets of "almost" identical images on your camera. Now comes the fun part in playing with the data you have captured so it gives you a much prettier picture. There are two pieces of free software available to help you process your images. Let's look at them briefly in turn:

Startrails: http://startrails.de/
This is a nice piece of software that allows you to add images together. As its name suggests, it enables you to add images together to produce star trail pictures. If your camera is static on a tripod (as yours currently is) the stars are moving across the sky due to the Earth's rotation. This means that between each of your exposures, the sky has moved. If you take enough images and put them into Startrails it will

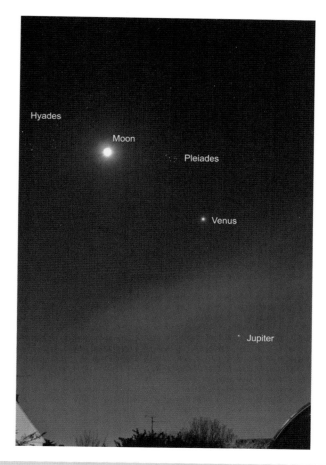

Fig. 13.4 A single un-driven image of the planets in the evening sky (Courtesy of the author)

show the difference in the stars position and you will see the movement quite easily. The use of JPG's suggested earlier would be better for this as they are smaller images and more easily handled by the Startrails software. To show you how it works, Fig. 13.5 is a 20 s image of Aquila taken with a DSLR from a dark sky site in Wales.

It doesn't look very exciting so far, does it? At the same time as this image was taken 17 images exactly the same were also captured. All the images were processed using Startrails software. The resulting image is shown in Fig. 13.6.

Now, that does look a little more inspiring, doesn't it? Startrails enables you to see the movement of the stars and produce star trails from the images you have already taken. The colors of the stars also show up very well in an image processed

Fig. 13.5 A single 20 s image of milky way (Courtesy of the author)

like this. Startrails will also produce an animation of the sky's movement throughout the period of capture and save it as an AVI file.

Processing Your Images Using Deep Sky Stacker

Now things get really exciting and will really bring out the Milky Way from your images. Using the very same 17 RAW images of Aquila shown above, they were then processed using Deep Sky Stacker. Unlike Startrails, DSS looks at the images and determines where the stars are. So when it processes the images, despite the fact that the stars have moved between exposures, and even the field of view having rotated, it registers the stars on top of one another and adds all the images together. As a result of this faint objects that you can hardly see on the original images are made much brighter as the data is added together. Figure 13.7 shows the result of stacking the same 17 RAW images of Aquila together.

Fig. 13.6 Multiple milky way images processed using startrails (Courtesy of the author)

Now you can clearly see the bright band of the Milky Way and the dark rift at the side of Aquila. If you look carefully you can also see the Coat Hanger in Vulpecula easily visible. I'm sure you will agree that this is not a bad result for a camera standing static on a tripod and not tracking the sky at all. It would certainly enable you to record the sky just as you can see it with the naked eye.

Using a Pocket Camera

Pleasing images can still be taken using small pocket cameras. Many pocket cameras have a night landscape mode built in. This focuses the camera at infinity and exposes the chip for about 2 s. This is enough to record most of the brighter naked eye stars. Small compact cameras sometimes have a Program or Night Landscape mode. This automatically sets a long exposure of a few seconds and focuses the camera at infinity. Some nice constellation images can be taken this way.

Fig. 13.7 Multiple milky way images processed using deep sky stacker (Courtesy of the author)

13.5 Astrophotography Through the Telescope

Moving onto taking images through a scope requires a bit more dedication and thought to the type of equipment you will need. The challenges and subsequent problems encountered increase dramatically once you start trying to image through a telescopes, but it can still be started without too much financial outlay.

- Background reading. As much information as possible.
- A telescope on any type of mount.
- A bright object, such as the Moon.

- Any type of camera. This would ideally be a DSLR, but it could also be a pocket camera. Some observers are even managing to take images of the Moon and the larger planets using the cameras built into mobile phones.

Mount

A sturdy mount is probably the single most important factor for good astrophotography. Get this right and it makes the rest just that little bit easier. If you've got a shaky mount, you will never get a good image despite how good the optics of your scope might be. Get the best mount that you can afford. Some extremely good Chinese built mounts are currently dominating the market. And it's not surprising, these give great value for money and are extremely well built.

Camera

The best results will always be obtained by DSLR's or dedicated CCD cameras. However these are expensive. Pocket cameras can be used to image the sky with some degree of success. Successful images have even been taken of the Moon and planets using the camera built into a mobile phone.

So whatever type of imaging equipment you have, get out and give it a go.

Mounting the Camera on the Telescope

You may find that if you use an inch and a quarter or two inch adapter to attach a camera to the telescope you may not reach focus properly. The camera will be held too far out and the focuser will not move in far enough to reach the focal point.

Fig. 13.8 Lunar image taken using a mobile phone (Courtesy of Dusko Novakovic)

Many of the telescopes available today have an inch and a quarter adapter which screws off the end of the focuser. The T-adapter that attaches your camera can then be screwed directly onto this thread. This will enable you to reach prime focus.

Bright objects, like the Moon can be imaged through a telescope with even the simplest form of camera. Some pleasing results can even be obtained by holding a mobile phone up to the eyepiece onto the focused image and taking a snap. This is imaging using a focal projection.

13.6   Imaging the Planets

Once you start to try taking images of the planets you soon find out the biggest problem in doing so. The planets are just too small to see any real detail. Even Jupiter, the largest of the planets only shows a few belts when used with a normal camera. To take detailed views of the planets, most imagers prefer you use webcams. This will, in most cases require a reasonably good tracking mount, although they can be imaged by letting them drift through the field of view. This will result in a lower number of frames and reduce the quality of the resulting image.

Webcams can be picked up relatively cheaply. The preferred type of webcam for imaging the Moon and planets is one containing a chip that is of the CCD type. Some webcams contain a CMOS chip and these used to be shunned due to their lower sensitivity. However CMOS chipped cameras are becoming harder to find brand new these days.

An overview of Webcam imaging is available on the Starizona Web site:
http://starizona.com/acb/ccd/software/soft_image_webcam.aspx

Software for Webcam Processing

AVI capture software:
Sharpcap:
http://www.sharpcap.co.uk/

Processing Images
AVI files can be processed using:

Registax Software:
http://www.astronomie.be/registax/
AviStack:
http://www.avistack.de/

13.7   Imaging Deep Sky Objects

Fainter objects like star clusters or faint nebulae will need much long exposures to reveal them properly. These can result in many hours exposure. Once you start down this road you inevitably start to encounter many more problems. No wonder many people regard astrophotography as a dark art.

Fig. 13.9 Long exposure auto-guided astrophotograph of the pleiades star cluster, showing nebulosity (Courtesy of the author)

In a similar way to the process in stacking raw files above, the same process is used to image faint objects through the telescope. You simply take a number of images, usually many minutes long, and stack them together. This adds the signal and reduces the noise to produce a much better image.

It is hoped that this quick primer will encourage you to give it a go. You really won't be disappointed. Once you do start, you will be gradually drawn well and truly into "The Dark Side", where auto-guiding, flat fielding and dark frames really do become part of your standard work flow.

13.8   Astrophotography Resources

Making Every Photo Count. A Beginners Guide To Astrophotography. (Book).

An extensive account of everything you will need to start out in astrophotography.

Richards, S. Chantonbury Observatory. www.nightskyimages.co.uk

Digital Astrophotography. A Guide to Capturing the Cosmos. (Book)

Seip, S. RookyNook.

A Guide to Astrophotography, With Digital SLR Cameras. (CD-ROM)

J. Lodriguss. www.astropix.com.

Useful Imaging Software

Startrails: Software for producing Startrails or animations from a series of un-driven star images. http://startrails.de/

Deep Sky Stacker (DSS). Software used for stacking a number of astronomical images to add the data together. http://deepskystacker.free.fr/english/index.html

IRIS. Image processing software specifically written for astronomical image processing. http://www.astrosurf.com/buil/index.htm

The GNU Image Manipulation Program (GIMP). Free image processing software for all platforms. http://www.gimp.org/

Part III

**Further
Information
Sources and
Community**

Chapter 14

Online Resources

14.1   Software

There are a wide variety of software programs that can, in many cases be freely downloaded from the Internet. The software highlighted here are generously supplied by their developers to download for free.

Stellarium

Planetarium Software suitable for Observing, with lots of add-ons.
  Richly graphical and very accurate. http://www.stellarium.org/

Computer Aided Astronomy (C2A)
Extremely accurate planetarium software, including faint stars and easy to update comets and asteroids.
  http://www.astrosurf.com/c2a/english/

Cartes Du Ciel
Planetarium software very similar to C2A.
  http://www.ap-i.net/skychart/index.php

Guide
Low cost software from Project Pluto.
  Highly detailed planetarium software.
  http://www.projectpluto.com/

D. Eagle, *From Casual Stargazer to Amateur Astronomer: How to Advance to the Next Level*, The Patrick Moore Practical Astronomy Series, DOI 10.1007/978-1-4614-8766-1_14, © Springer Science+Business Media New York 2014

Graphdark
Visibility of objects and Moon phases from your location can be calculated with
this program written by Richard Fleet for Windows based PC's.
   http://home.clara.net/rfleet/graphdark/gdkindex.html

XEphem
Unix-based planetarium software. Takes a bit of setting up, but very useful once
mastered. Low cost upgrade available with more extensive databases.
   http://clearskyinstitute.com/xephem/

   An extensive list of astronomy related software for all platforms is available on
this Web page:
   http://www.midnightkite.com/index.aspx?AID=0&URL=Software

14.2 Online Citizen Astronomy

These days even someone who cannot get out and observe, for whatever reason,
can still make a useful contribution to astronomy. Professional astronomers are
having problems keeping up with the influx and interpreting the data. Citizen
Science has become a very useful aid in getting this data processed. A number of
different online databases have been set up so that people with time on their hands
can download particular parts of the data. They can look at images and interpret the
details themselves, or report anything that might be unusual. This has already
enabled a few people sitting in their homes to make discoveries that professional
astronomers would have otherwise missed or not discovered for many years.

Zooniverse
This huge suite of Web resources enables you to look at different types of data.
This project covers a wide and varied range of data, not all of it astronomical.
Planet Four is among their best ventures, a site devoted to the Red Planet that is
meant to identify and categorize features on the surface of Mars. There are also a
number of other projects covering such diverse subjects as Climate, Humanities,
Nature and Health. The concept is that you sign-up to a project of interest. You are
assigned a set of data to download and analyze. When you've finished you submit
your results. The hundreds of thousands of people now involved in the projects
ensures great results are achieved in order to achieve the best determination from
all that data.
   www.zooniverse.org

Galaxy Zoo
Looking at images of galaxies and making a determination of their classification

Moon Zoo
Identifying and characterizing lunar surface features using images from the Lunar
Reconnaissance Orbiter.

Solar Stormwatch
Looking at images of the Sun and spotting eruptions on the Solar Surface and track them towards Earth for possible Borealis displays.

Planethunters
Looking for Exoplanets: Remote planets around distant stars using light curves from Kepler data.

Space Warps
Looking for Gravitational Lenses.

The Milky Way Project
Looking at Spitzer and Herschel infrared data to interpret star formation within the Milky Way.

The Andromeda Project
Spotting and classifying star clusters in M31 from HST images. The first stage of the project is completed, with more data to be released in 2013. http://www.andromedaproject.org/

GAIA
Data from the billion star 3D GAIA survey mission is soon to be made available. So keep a look out to take part in this exciting citizen science project.
    http://www.esa.int/Our_Activities/Space_Science/Gaia_overview

    There is also a wealth of data and images freely available for downloading:

Sky-Map.org
Searchable online star atlas.
www.sky-map.org
The Sloan Digital Sky Survey
The most up to date digital sky survey. New results and images are being added
    all the time.
http://www.sdss.org/
The STScI Digitized Sky Survey
Download up to half a degree sections of the digitized Palomar Sky Survey.
    Monochrome Red and Blue sensitive plates only.
http://archive.stsci.edu/cgi-bin/dss_form
SIMBAD
Comprehensive online database of celestial objects extensively used by profes-
    sional astronomers. http://simbad.u-strasbg.fr/simbad/
ALADIN
Interactive Sky Atlas used to visualize digitized astronomical images and super-
    impose astronomical catalogues and databases.
http://aladin.u-strasbg.fr/aladin.gml
Hubble Space Telescope Images
Hubble Heritage Project: http://heritage.stsci.edu/index.html
Hubble Site Gallery: http://hubblesite.org/gallery/

emote Telescopes

.d your skies are mostly cloudy? Do you want to take an image of
that is too far south to be seen? Then help is at hand. There are a
number of remote observatories that have been set up all over the world that can
be used by anyone with a computer and an Internet connection. Most of them
require a fee for use, but many are available for free, especially for educational
establishments.

iTelescope.net
A network of Internet connected telescopes that charge for their use.
   http://www.itelescope.net/contact-us

The Bradford Robotic Telescope
A collection of telescopes in Tenerife in The Canary Islands set up by the University
of Bradford. These are available for all to use for a nominal sum.
   http://www.telescope.org/

The Foulkes Telescope
These two telescopes, North in Hawaii, South in Australia are freely available for
use in education. Schools and associated local astronomical societies can use the
scopes and download the resulting images.
   http://www.faulkes-telescope.com

14.4   Astronomy Podcasts

There are a number of astronomical related audio podcasts freely available on the
Web. Here are a few of the most popular ones. Free registration is sometimes
required.

NASA
http://www.nasa.gov/multimedia/podcasting/index.html
Astronomy Magazine
http://www.astronomy.com/Multimedia/Podcasts.aspx
Awesome Astronomy
http://awesomeastronomy.com/home/
Jodcast
Podcast from the staff at Jodrell Bank, England.
http://www.jodcast.net/
365 Days of Astronomy
http://cosmoquest.org/blog/365daysofastronomy/
Slacker Astronomy
https://www.slackerastronomy.org/category/audio-podcasts/

14.5 Mobile Phone/Tablet Apps

Technology has certainly come a long way in the past few years. With advanced smart phones, you can quite literally carry the universe around in your pocket. All that information as well as the Internet is just there at your finger-tips wherever you may be (providing there is a mobile internet network where you are, of course).

There are many different astronomical related apps available to download on all the different platforms.

Popular titles include Google Sky, Sky Safari and Sky Walk. Many of these are free or can be purchased at very reasonable prices.

There are many more titles covering most aspects of astronomy.

They vary enormously in their usefulness and functionality, ranging from extremely good to totally useless.

Look in your application catalogue for your device and your particular operating system.

Listen to what other people use and reviews online. You will very soon end up with lots of apps on your tablet or smart phone that will becomes invaluable weapons in your astronomical armory.

14.6 Meeting Other Observers

There is nothing like sharing your experiences with other people. Meeting others with the same interest can only encourage you to observe more.

Organizations that help people get together to share our hobby can be organized at different levels:

Fig. 14.1 With the right apps, smart phones and tablets become excellent observing tools (Courtesy of the author)

14.7   Local Astronomical Clubs/Societies

Observing the sky can be a very lonely hobby. Standing out in the freezing cold in the middle of the night, it is quite easy to feel that you are the only person mad enough want to be out there out of choice. Meeting up with other enthusiasts in your local club can really help your development as an observer. You see first-hand what other people are able to achieve, which will, in turn, encourage you to do more. Many societies build up a great sense of camaraderie, especially when out observing in a group. Nothing beats the combined cries of "Whooo!" when something like a bright meteor flashing across the sky is witnessed by many of those present. The experience within a local group will vary enormously, from beginners who are new to the hobby, to active observers who gather data diligently. There are also quite a number of armchair astronomers who rarely go out observing themselves, but just love to hear about and talk to people who are interested in their favorite subject.

Activities arranged by local clubs are as many and various as the people they are comprised from. They frequently hold more formal meetings once or twice a month. How formal, or not those meetings actually are depends on the group. They arrange talks by eminent people in their fields as well as talks given by their own members showing what has been recently observed. The best societies organize observing sessions, especially when certain celestial events are expected, like an

Fig. 14.2 An observing day organized by bedford astronomical society for the 2008 Venus transit (Courtesy of the author)

eclipse. Many have a library of books, or a collection of small instruments, which can be borrowed by their members. Many publish newsletters that are distributed to members, letting them know what celestial events are going to happen each month and keep them informed of the latest astronomical news. They are also a great place to look through other people's telescopes if you are in the market to buy a new one or upgrade. For a nominal fee, the benefits available from a local club are enormous.

If you do find a club you like, get involved! It will become even more rewarding experience. Perhaps you find that your favorite club doesn't organize a type of event that you and others would enjoy. Make suggestions and offer your services to run one or two of these events yourself. Any self-respecting club should snap the hand off a willing volunteer. A society can only become richer from fresh ideas. You might be that breath of fresh air ready to boost the clubs activities. A club is only as good as the effort its members put into it.

In some cases you might even fail to find a club near enough to your area? So what's stopping you from starting one yourself? It can be challenging and demanding running an organization like this single-handedly in the early days, but the rewards from such an endeavor make it well worthwhile in the long run. As you attract more members, a "Committee" of willing volunteers will very soon form. The workload then starts to spread out somewhat as other people take on specific roles. All too soon you find you will find that you have a fully-fledged astronomical society with all the benefits it brings.

14.8 National and International Organizations

There are a number of organizations scattered around the world which act as umbrella organizations. Many of them encourage observations of various objects and gather the results. As a result very useful observational databases are quickly established. These can be used to add to our understanding of the universe and how it all works. If you do get involved, make sure that you try to regularly submit observations. If you attend some of their meetings on a regular basis you will soon get to know other people. When you start to regularly meet up with people at organized events such as this, it can soon become a very sociable experience as well.

International
The largest organization is undoubtedly the International Astronomical Union. This professional organization, founded in 1919 has a mission statement to promote and safeguard the science of astronomy in all its aspect through international cooperation. It's committee has the final say in all things astronomical.
    http://www.iau.org

United States
The American Astronomical Society has a mission statement to enhance and share humanity's scientific understanding of the universe. This society is mainly aimed at the professional observer.

http://www.aas.org

The American Association of Variable Star Observers (AAVSO) encourages the observation of variable stars internationally. It collates observers data and over the years has built up an extensive archive of observations.

http://www.aavso.org

The Astronomical League is an umbrella organization for local astronomical societies in the United States. Their Web site has a number of observing programs that can be used to encourage people to get out and observe for themselves. There are a wide range of programs there to keep the keen observer busy for many years. Many of the programs can be downloaded in pdf format and printed for later use. You do not have to be a member of the Astronomical League in order to access many of these.

http://www.astroleague.org

United Kingdom

The Royal Astronomical Society encourages and promotes the study of astronomy, solar-system science, geophysics and closely related branches of science. Mainly geared towards the professional observer, although anyone can be elected as a fellow, not only from the UK.

http://www.ras.org.uk

The British Astronomical Association (BAA) encourages amateur astronomers to observe and submit their results. The BAA is divided into different observing sections. These observations are then compiled and added to further the wealth of knowledge building up a much broader picture. The BAA has members all over the world. Produce a yearly handbook of astronomical data.

http://britastro.org/baa/index.php

The Astronomer Magazine is a subscribed magazine, both printed and online that cascades the latest astronomical news. The magazine alerts active observers so they can get out observing and start to send back observations of new discoveries as quickly as possible.

http://www.theastronomer.org

The Society for Popular Astronomy. A UK based society that encourages more relaxed observers and is particularly encouraging for beginners and young astronomers. Also has separate observing sections to encourage observing.

http://www.popastro.com/

In the UK, a similar role to the Astronomical League is achieved by the Federation of Astronomical Societies. The main considerations of this organization are to meet the needs of local astronomical societies, clubs and associations. They publish a yearly sky calendar and a handbook listing useful information and speakers willing to visit UK societies. They have an extensive list of astronomical societies in the UK.

http://www.fedastro.org.uk

Webb Deep-Sky Society encourages observations and imaging of double stars and deep-sky objects. They have an international membership and publish a number of atlases and useful observing handbooks related to deep-sky observing.

http://www.webbdeepsky.com

Rest of the World

The Royal Astronomical Society of Canada. Produces a bi-monthly journal and an annual Observers Handbook which highlights celestial events for the year. http://rasc.ca/

The Astronomical Society of Australia is the organization for professional astronomers in Australia.

http://asa.astronomy.org.au

The Astronomical Society of the Pacific. Advancing Science literacy through Astronomy.

http://www.astrosociety.org/

In Australia the Astronomical Society of Australia Web page has a list of Australian amateur astronomical societies.

http://www.astronomy.org.au/ngn/engine.php?SID=1000022&AID=100136

Astronomers Without Borders

An international society who's aims are to foster understanding and goodwill across national and cultural boundaries by creating relationships through the universal appeal of astronomy.

http://www.astronomerswithoutborders.org/

14.9   Online Discussions and Resources

Thankfully, with all the online resources now available it is still possible to take part in astronomical activities and discussions even if you don't have a telescope. As well as keeping you up to date with the latest discoveries and sky diaries, some of the Web sites now available will also enable you to download valuable data for you to carry out your own analysis. Therefore you can still make a valuable contribution to astronomical research no matter what your circumstances. Watch out though as the Internet can be a source of unlimited misinformation as well.

Facebook, Twitter and other popular social Web pages have many groups dedicated to astronomy and related subjects within them. Quality of the discussions can vary enormously.

Night Sky Observer. Regular updates on news and astronomical events.
http://www.nightskyobserver.com
Stargazers Lounge (SGL). A family friendly online astronomy forum.
http://stargazerslounge.com
Inside Astronomy. Astronomical online forum.
http://www.insideastronomy.com
The Astronomy Shed. Astronomical online forum.
http://www.astronomyshed.co.uk/forum/index.php
Yahoo! Discussion Groups. There a variety of different astronomical observing and
    imaging discussion groups within Yahoo!

http://groups.yahoo.com/

Astronomy Forum. An Australian based forum, although it caters for the international community as well.

http://www.astronomyforum.net

Ice In Space. An astronomy forum dedicated to promoting astronomy in the southern hemisphere.

http://www.iceinspace/com.au

14.10 Astronomy Conventions

Many of the umbrella organizations listed previously organize regular get-togethers with eminent speakers, trade displays and plenty of opportunity for socializing. The many commercial vendors on show often have special deals for the day, making it well worthwhile. For the amateur who really wants to get involved with in-depth scientific debate, there are also quite a number of professional conventions each year on varied subjects. See the relevant professional organizations for more details of these.

North East Astronomy Forum.

Held at Rockland Community College, Suffern, New York.

http://www.rocklandastronomy.com/NEAF/index.html

Pacific Astronomy and Telescope Show.

Held at the Pasadena Convention Center in Southern California.

http://www.rtmcastronomyexpo.org/pats/index.php

European Astrofest

Organized by Astronomy Now Magazine. Kensington, London, England.

http://europeanastrofest.com

ATT

Organized by the Walter-Hohmann Observatory this convention, billed as "Europe's Biggest Astronomy Fair" is held in Essen in Germany.

http://www.att-boerse.de/index_uk.html

International Astronomy Show.

Organized by UK Astronomers. Its first meeting was held in Leamington Spa, Nr. Birmingham, UK in May 2013.

http://www.international-astronomy-show.com

National Australian Convention of Amateur Astronomers.

This roaming convention is hosted bi-annually by a different Australian astronomical organization.

http://www.nacaa.org.au/

South Pacific Star Party.

Held by the Astronomical Society of South Wales.

http://www.asnsw.com/

Starfest.

Canada's largest annual observing convention & Star Party.

http://www.nyaa.ca/

14.11   Remote Dark Sky Observing. Going it Alone

If you live in a town you will want to get well way form the glow of the streetlights. Many observers these days do arm themselves with equipment that is small, light and fairly portable to make the most of this type of portable observing. Finding a dark and fairly safe observing location is the main priority. Make sure you are not trespassing on private land. Owners often force intruders off their land forcibly, so it's not worth the effort. If you do find a good location, try to get permission from the person who owns it. Not only will you have permission, but it's another person who will know where you are, adding much to the safety aspect.

Make sure that if you do drive out for some miles, that you are well prepared. Wear lots of layers of clothes. Even in summer it can get quite cold on clear nights. Once you get cold, it's game over. Make sure you have a printed check-list of things to bring. There's nothing more disheartening after driving for miles to do your observing to arrive and find that you have forgotten an essential piece of equipment. Also check that your equipment works before you go. It's much easier to repair at home under a light than it is in the dark using a dim torch or in the car headlights.

Safety is the key. Make sure that you are not putting yourself at risk by being out alone at night. Don't forget it's your personal safety that matters most, not forgetting you will be out with some expensive equipment. Try to organize a field trip with another observer and try not to go alone if at all possible. It's always much safer when you are observing in numbers.

14.12   Remote Dark Sky Observing. Star Parties

Gatherings of astronomers at dark sites to enjoy observing and imaging is becoming increasingly popular. It adds much to the safety aspect of observing in a dark location enjoying your observing. Rather than being alone you are surrounded by plenty of fellow astronomers. Observers of all levels of ability and knowledge attend star parties, bringing with them a wide range of experience, telescopes and other equipment. These gatherings are usually held at sites well known for their dark skies, well away from large towns and problems of light pollution. A number of well-established events are held all over the world. Due to the rising popularity of astronomy as a hobby and the fact that many of these sell out really quickly, many more events are now starting to emerge. Many events have grown to hundreds of attendees every year. This makes them extremely enjoyable, highly informative and extremely sociable.

The general format of the event will depend on the facilities at the site and the way it is organized. If camping is available, people pitch their tents or mobile homes setting up their equipment on their own pitch. Some astronomical societies block book a number of pitches. Their accommodation pitched on selected sites, leaving a number of communal pitches on which to set up their equipment. This makes the observing very sociable.

Fig. 14.3 The author ready for an evening of imaging at a star party in Kelling, Norfolk, UK (Courtesy of the author)

Many of the larger events often have invited speakers giving talks during the main event day(s). Commercial vendors are often present selling astronomical equipment and accessories. Many attendees bring along their second-hand equipment and display it for sale. Most star parties are extremely popular events. You often have to book your space more than a year in advance to ensure success. So if you are thinking of booking as star party for next year, book early!

14.13 Basic Rules for Star Parties

There are some basic rules of etiquette that should be followed if you do attend a star party. Only use red light sources at night and keep the light facing down. After all, the astronomers have probably travelled a long way to this dark sky site, don't spoil it! To reduce the problem of headlamps interfering with the darkness, car movements are not usually permitted after nightfall, so please arrange to arrive before it gets dark. After dark, avoid using any car lights, exterior and interior, whilst on site. Do not keep locking and unlocking your car that make the lights flash. It is also a good idea to switch off, or cover any unwanted internal lights so that when you do need a small amount of light from inside the car they do not swamp the observing area with light.

Computer screens are very bright, especially at night. Please dim the brightness if possible and shield the screen from general view. Attach a sheet of red cellophane sheet across the front and use night vision mode if possible. Keep talking/music/noise on site as low as possible to avoid disturbing anyone. Don't forget, not everyone stays up late observing. Don't forget those who will still be asleep the next morning after a late night of observing.

Although children are usually very welcome, if you bring them along please remember the field is dark. There are some very valuable bits of equipment set up around the site. Please keep them under control and as quiet as possible. Well-behaved dogs are welcome, but when they are not, they can be a real nuisance. Especially if they are running loose in the dark or leaving presents on the ground for someone to tread in.

Lastly, keep to the main thoroughfares. Don't trample the natural environment and disturb the wildlife. Most of the sites used for star parties are usually rich in wildlife and very pleasant places to stay. Please keep them that way. Dispose of your rubbish properly, or better still, take it home and dispose of it there.

14.14   Some of the Biggest and Most Popular Star Parties

United States
Texas Star Party.
  Organized by Austin Astronomical Society. This event is held at Fort Davis in Texas in the US. Registration required.
  http://www.Texasstarparty.org
  Winter Star Party.
  Organized by The Southern Cross Astronomical Society. This event is held in The Florida Keys in the US. Ticket Only.
  http://www.scas.org/wsp.html
  Stellafane Convention. Organized by Springfield Telescope Makers. This event is held in Springfield, Vermont in the US.
  http://www.stellafane.org

United Kingdom
UK Spring Equinox Sky Camp.
  Organized by Norwich Astronomical Society. This event is held at Kelling Heath, Norfolk in the UK.
  http://www.starparty.org
  UK Autumn Equinox Sky Camp.

Organized by Loughton Astronomical Society. This event is held at Kelling
Heath, Norfolk in the UK.

http://www.starparty.org.uk

Stargazers Lounge.

Their SGL events in the UK have proved very popular. Open to all members of
the online astronomy forum.

http://stargazerslounge.com

Rest of the World

Teleskoptreffen

Held on Lake Gederner in Bavaria, Germany.

http://www.teleskoptreffen.de/

Starfest.

Organized by the North York Astronomical Society in Canada.

http://www.nyaa.ca/index.php?page=/sf13/sf.home13

A number of other Canadian astronomical Star Parties are listed by the Royal
Astronomical Society of Canada:

http://rasc.ca/star-parties-and-events

Queensland Astrofest.

Held at Lions Camp Duckadang, Linville, Queensland, Australia.

2 h north of Brisbane.

http://www.qldastrofest.org.au/

Worldwide Star Parties.

A huge number of worldwide events are listed by Amsky. Some entries are out
of date, but fortunately many of the links still lead to active sites:

http://www.amsky.com/calendar/events/

Chapter 15

Star Maps, Sources of Information, Publications and Further Reading

15.1   Star Maps

With the advent of the computer and the availability of fantastically accurate planetarium programs, the printed atlas has largely been replaced these days. For those who enjoy the printed article, or need something convenient for use at the telescope, the following are recommended. Many of the larger format maps are now out of print, so are now becoming harder to find.

Don't forget the humble Planisphere. Many types are available for different latitudes. They show the whole sky at one viewing for easy reference.

Collins Gem. Stars (or The Night Sky). Ridpath, I, Tirion, W. Collins. Handy pocket-sized book.

Cambridge Sky Atlas. Tirion, W. Cambridge University Press. Size 12×9 in.. Plots stars to magnitude 6.5 and 900 deep-sky objects.

Sky & Telescope Pocket Sky Atlas. Sinnot, R.W. Sky & Telescope. Small format book 6×9 in.. Very useful for use at the telescope. Plots 30,000 stars to Magnitude 7.6 and 1,500 deep-sky objects.

The Monthly Sky Guide. Ridpath, I. & Tirion W. Cambridge University Press.

Norton's Star Atlas. Ridpath, I. Longman Press. Now in its 20th edition, this classic book is not only a star atlas, but also contains a wealth of astronomical information, facts and figures.

Sky Atlas 2000. Tirion W. Cambridge University Press. 1981. Larger format 12.7×16.7 in., plots 90,000 stars and 2,700 deep-sky objects.

D. Eagle, *From Casual Stargazer to Amateur Astronomer: How to Advance to the Next Level*, The Patrick Moore Practical Astronomy Series, DOI 10.1007/978-1-4614-8766-1_15, © Springer Science+Business Media New York 2014

Fig. 15.1 A planisphere designed for use at 51.5° North (Courtesy of the author)

Fig. 15.2 A selection of recommended star maps (Courtesy of the author)

Uranometria 2000. Tirion, Rappaport and Lovi. Willman-Bell. Heavy books so not quite so easy to use while out observing. Two editions, covering northern and Southern Hemisphere. Plots 280,035 stars and 30,000 deep-sky objects.

The Great Atlas of the Sky. Brych. Very large format loose-leaf maps 24 × 17 in.. Indoor use only, but do include a plastic sleeve for outdoor use. Plots 2,430,768 stars to magnitude 12, and 70,000 deep-sky objects. Now out of print. http://www. greatskyatlas.com/

TriAtlas Project. Sets of detailed star maps in a PDF format for free download. Three versions of increasing detail. Up to 9th, 11th or 13th magnitude stars. http:// www.uv.es/jrtorres/index.html

15.2 Other Publications

Data Book of Astronomy. Moore, P. & Rees R. Cambridge University Press. 2011. Facts and figures and a vast wealth of information of solar system, deep sky and the constellations.

Cosmic Challenge. Philip S Harrington. Cambridge University Press. 2012.

Turn Left At Orion. Consolmagno, G and Davis D.M. Cambridge University Press. 2011.

NightWatch. Dickinson, T. A&C, Black. 2006.

Illustrated Guide to Astronomical Wonders. Thompson, R.B. & Thompson, B.F. O'Reilly. 2007.

The Observers Year. Patrick Moore. Springer. 1998.

The Observational Amateur Astronomer. Patrick Moore (ed.) Springer 1995.

A Complete Manual of Amateur Astronomy. P. Sherrod & T.L. Koed. Dover. 1981.

Night Scenes. Annual sky diary for events in the coming year based in the UK. www.astrospace.co.uk

The following manuals are now a little out of date, but do contain some useful information on general observing techniques.

The Sky – A Users Guide. David H Levy. Cambridge University Press. 1991.

Sky Watchers Handbook. James Muirden. W. H. Freeman. 1993.

Astronomy Hacks. Tools and Tips for Observing the Night Sky. R.B. & B.F. Thompson. O'Reilly. 2005.

The Cambridge Encyclopedia of Amateur Astronomy. M.E. Bakich. Cambridge University Press. 2003.

15.3 Places of Astronomical Interest

There are many places of astronomical interest around the globe, from major research centers to small town observatories. Wherever you are in the world there will usually be somewhere to visit near you. A quick search for observatory and your location will soon produce some gems worthwhile visiting.

Below are listed some of the major locations around the world. These include links to other Web sites that will lead you to the information you will need to find one located reasonably close to you.

Observatories and Planetariums

These representations of the night sky range from large scale national facilities that can accommodate hundreds of people to smaller one-man portable visiting planetaria that can only accommodate a small number of people at a time. Many of the major universities majoring in physics have planetariums or observatories that can be visited on request. Many also organize open days or evenings and welcome visitors. Some observatories are more difficult to contact and visit. Listed below are some facilities that are welcoming.

The International Planetarium Society
This international organization has a comprehensive list of planetaria around the
    world: http://www.aplf-planetariums.org/en/index.php
Wikipedia also have a list of major planetariums located across the globe:
http://en.wikipedia.org/wiki/List_of_planetariums

Australian Planetaria and places of interest
A list of planetaria and observatories in Australia is listed on the Australian
    Observatories Web site:
http://www.astronomy.org.au/ngn/engine.php?SID=1000014
Another Australian source is here:
http://www.quasarastronomy.com.au/places.htm

The British Association of Planetaria
An umbrella organization for British Planetaria. A full listing of their members
    on their Web site: http://www.planetarium.org.uk/planetaria.asp

United States

Hayden Planetarium.
Run by the American Museum of Natural History in New York.
http://www.haydenplanetarium.org/index.php
Lowell Observatory.
Historic site at Flagstaff, Arizona, set up by Percival Lowell in his quest for
    Martian life and the site of Clyde Tombaugh's discovery of Pluto.
http://www.lowell.edu
Palomar Observatory.
This historic observatory located in North San Diego County, California, United
    States. Instruments include the 200 in. Hale Reflector and the 48 in. Oschin
    Schmidt telescope.
http://www.astro.caltech.edu/palomar
Kitt Peak Observatory.
The world's largest collection of optical telescopes located 1.5 h from Tucson in
    the Sonoran Desert, US.
http://www.noao.edu/outreach/kpoutreach.html
Yerkes Observatory.
Located in Williams Bay, Wisconsin, US and run by the University of Chicago.
http://astro.uchicago.edu/yerkes/

Arecibo Observatory.

The iconic radio telescope in Puerto Rico, run by the National Science Foundation. http://www.naic.edu/

Mauna Kea.

The site is maintained and run by the Institute For Astronomy at the University of Hawaii. There are a number of international observatories based here close to the summit of Hawaii.

http://www.ifa.hawaii.edu/mko/

United Kingdom

The National Maritime Museum and Royal Observatory.

The home of The Prime Meridian of the World at Greenwich, London, England.

Entrance to the Astronomy Centre is free, but other attractions such as Flamsteed's House, The Meridian Courtyard and the Planetarium are chargeable.

http://www.rmg.co.uk/royal-observatory

Kielder Observatory.

Located in dark skies in The Kielder Forest, Northumberland, England, just on the Scottish border. A number of telescopes can be used by paying guests.

http://www.kielderobservatory.org

Jodrell Bank.

The iconic radio telescope in Cheshire, England.

Operated by the University of Manchester.

http://www.jodrellbank.net/

Observatory Science Centre.

Located at the old Greenwich Observatory site in Hertsmonceaux, East Sussex, in Southern England. Site open most of the year, organizing observing events, courses and workshops.

http://www.the-observatory.org

The National Space Centre.

Family oriented visitors center in Leicester in the UK, with hands-on exhibits of space and its exploration.

http://www.spacecentre.co.uk/

Norman Lockyer Observatory.

Run by the Norman Lockyer Observatory Society. Located in Sidmouth, Devon, England.

http://www.normanlockyer.com/index.html

Scottish Dark Sky Observatory.

Located in Ayrshire, Scotland. They also run a mobile planetarium.

http://www.scottishdarkskyobservatory.co.uk/

Australasia

Parkes Observatory.

The iconic radio astronomical observatory in Australia, run by CSIRO, Australia's national space agency.

http://www.parkes.atnf.csiro.au/
Siding Spring Observatory.
Houses Australia's two largest optical instruments, located 500 km NW of
   Sydney.
Operated by the Australian National University.
http://rsaa.anu.edu.au/observatories/siding-spring-observatory
Mount John Observatory.
Maintained and run by the University of Canterbury. Located on South Island in
   New Zealand. They do not appear to have a visitor's center at the moment.
http://www.phys.canterbury.ac.nz/research/mt_john/
Aukland Observatory and Planetarium.
http://www.stardome.org.nz/
Carter Observatory.
Located in Wellington, New Zealand.
http://www.carterobservatory.org/
A list of Australian Observatories is listed on this Web site from the Astronomical
   Society of Australia.
http://asa.astronomy.org.au/observatories.html

Europe

Tenerife, Canary Islands.
There are major multi-national observatories on the island.
http://www.iac.es/?lang=en

15.4   Astronomical Holidays

There are a number of travel agents who specialize in astronomical holidays and
tours, ranging from Northern Lights Tours, Meteor watches, Cruises and Total
Solar Eclipses. A quick search on the Web will soon identify one that serves your
country and organizes tours that appeal to you.

For those who do not want full tours and would like a holiday where they can
do astronomy but remain self-sufficient there are a number of options available.
The following places are permanent setups who cater specifically for people who
would like to enjoy a holiday or break with telescopes and other equipment available to use that will enable them to enjoy some astronomy at the same time.

United States
Arizona Sky Village.
   A holiday resort set up by the renowned astrophotographer Jack Newton.
Actively maintaining their dark skies in Arizona.
   Both accommodation and telescopes are available for rent.
   http://www.arizonaskyvillage.co.uk

Europe
Centro de Observação Astronómica no Algarve (COAA).
Located in The Algarve in Portugal, this resort is run on a Bed & Breakfast basis, with observatories and telescopes available free of charge for staying guests.
http://www.coaa.co.uk

Les Granges, France
Traditional Provence-style farmhouse for rent with all astronomical telescopes and imaging equipment available to use.
http://www.sunstarfrance.com

United Kingdom
AstroAdventures.
Located in North Devon, England, this resort is run on a Bed & Breakfast basis, with observatories and telescopes available free of charge if you are staying.
http://www.astroadventures.co.uk

Galloway Astronomy Centre
Bed & Breakfast accommodation located in The Dark Sky Park area in the Galloway Forest National Park, Dumfries, Scotland. A range of astronomical equipment is available.
http://www.gallowayastro.com/index.htm

15.5 Further Reading/References

No one resource will capture everything of interest. There are a huge number of publications with glossy images and up to date information available right across the globe.
Magazines.

United States

Astronomy Magazine.
Colorful, glossy, monthly US news stand publication.
http://www.astronomy.com/
Sky & Telescope. Monthly US news stand publication.
www.skyandtelescope.com
There is also an Australian Version of the magazine.
http://www.austskyandtel.com.au

United Kingdom.

Astronomy Now.
A colorful glossy monthly UK based news stand publication.
http://www.astronomynow.com/

Sky at Night.
Monthly UK news stand publication.
http://www.skyatnightmagazine.com/
Astronomy and Space.
Colorful glossy Ireland based magazine available by online subscription.
http://www.astronomy.ie
Rest of the World
SkyNews.
Colorful glossy Canadian based news stand magazine.
http://www.skynews.ca
Space.com.
Latest astronomy and space related news.
http://www.space.com
Amateur Astronomy.
Subscription based magazine.
Written by amateur astronomers, for amateur astronomers.
www.amateurastronomy.com
NASA's picture of the day. Features the best astronomical image each day.
http://apod.nasa.gov/apod/astropix.html
International Astronomical Union Central Bureau For Astronomical Telegrams.
Issues official announcements about the most recent astronomical discoveries. A
    subscription is needed to receive the most up to date announcements.
http://www.cbat.eps.harvard.edu/Headlines.html
The Astronomer Magazine. Alerts advanced amateurs (subscription required) of
    the latest astronomical discoveries and possible upcoming observing
    opportunities.
http://www.theastronomer.org/
Night Sky Network. A US coalition of amateur astronomy clubs bringing the
    science, technology, and inspiration of NASA's missions to the general pub-
    lic. http://nightsky.jpl.nasa.gov/
Astronomers Without Borders.
An international society who's aims are to foster understanding and goodwill
    across national and cultural boundaries by creating relationships through the
    universal appeal of astronomy. http://www.astronomerswithoutborders.org/
Astro Grid. Virtual Observatory. Software for Astronomers and data centers.
http://www.astrogrid.org/theVO.html
CalSky. Comprehensive Web pages that can be used to predict the visibility of
    astronomical objects, satellites, planets and Moons from your chosen observ-
    ing location. It will produce an astronomical calendar that can be imported
    into popular calendar software. A donation is appreciated if you do find the
    Web site useful.
http://www.Calsky.com
US Naval Observatory. Publish Moon rise and set times as well as other astro-
    nomical phenomena.

http://www.usno.navy.mil/USNO/astronomical-applications/data-services/data-services

RASC Observers Handbook. Detailed publication of the years celestial events. http://www.rasc.ca/observers-handbook

Night Scenes. Annual book outlining upcoming celestial events each month. Astrospace Publications. www.astrospace.co.uk

Cloudy Nights Telescope Reviews.
Have you been spurred on and now fancy getting yourself a new shiny piece of kit? Check this Web site to see if they have reviewed it already and how well it faired. www.cloudynights.com

Starware: The Amateur Astronomers Guide to Choosing, Buying and Using Telescopes and Accessories. Philip S. Harrington. 2007. John Wiley & Sons.

Bad Astronomy.
Think you know all the facts? Check here first to make sure you've not been misled. A great review of all the misconceptions and myths that raise their ugly head time and time again.
http://www.badastronomy.com/bad/misc/index.html
Phil also has a book published:
Bad Astronomy: Misconceptions and Misuses Revealed, from Astrology to the Moon Landing Hoax. Plait, P. 2002.

Space Weather.
News and observations about solar system objects, Solar Activity, aurora and meteors.
http://spaceweather.com/

Burnham's Celestial Handbook. Burnham Jr., R. (3 Volumes). New York: Dover, 1978. A lot of the information is now quite outdated, but even today is still an extremely useful reference work.

Cosmic Challenge. The Ultimate Observing List For Amateurs. Philip S. Harrington. Cambridge University Press. 2012. A selection of solar system and deep sky objects to challenge the observer, ranging from naked eye to very large telescopes.

15.6 Space Travel and Exploration

Sources of manned and unmanned spaceflight can be obtained from the appropriate space agency or society.

NASA: http://www.nasa.gov/
The European Space Agency: http://www.esa.int
Canadian Space Agency: http://www.asc-csa.gc.ca/index.html

Russian Federal Space Agency: http://www.roscosmos.ru/
Japanese Aerospace Exploration Agency: http://www.jaxa.jp/index_e.html
Indian Space Research Organization: http://www.isro.org/
Chinese National Space Administration: http://www.cnsa.gov.cn/n1081/index.
   html
Iranian Space Agency: http://www.isa.ir/index.php

Space Exploration Societies
There are a number of associations that cater for space travel enthusiasts.
The National Space Society. Produce the quarterly Ad Astra magazine.
http://www.nss.org/about/
The American Institute of Aeronautics and Astronautics.
Formed from the 1963 merger of The American Rocket Society and the Institute
   of Aerospace Science.
https://www.aiaa.org/default.aspx
A list of relevant events and conferences are available on their calendar:
https://www.aiaa.org/EventsCalendar.aspx?id=79
The British Interplanetary Society.
http://www.bis-space.com/
The National Space Society of Australia.
http://www.nssa.com.au/
The KiwiSpace Foundation.
http://www.kiwispace.org.nz/display/PORTAL/About+KiwiSpace

15.7   Light Pollution and Organizations

Light pollution has encroached onto our hobby over the years blotting out all but
the brightest objects from many people's night skies. Many amateur astronomers
now have to travel some distance to dark sky observing sites in order to carry out
their hobby. We all need lighting at night but sometimes any benefit is outweighed
by its detrimental effect on the night sky.

Many designs of light are far too bright for their use, and spread light upwards
and outwards as well as downwards. Not only lighting up our streets, but the sky as
well. If all the streetlights in the world were of the correct voltage for the job and
directed properly to where the light was actually required, the vast majority of light
pollution currently experienced by the observer just would not occur.

Do you have a street light that prevents you from observing? Or does a neighbor's security light have a similar effect? What do you do if a neighbors light is
interfering with your observing? Above all else, be nice to them. They may not even
be aware of the impact it's having.

If you are troubled by a streetlight shining into your garden, your first course of
action should be to contact your local lighting authority. Speak to them nicely to
see if there is anything that can be done to reduce the nuisance.

If you do feel that nothing can be done, luckily we are not alone in tackling these issues. There are organizations out there trying to make lighting more effective and less intrusive and can fight on our behalf to reduce light pollution and the affect it has on observing.

International Dark Sky Association.

An international organization campaigning for correct and appropriate lighting to preserve access to the night sky. http://www.darksky.org/

The Dark Sky Association have named a number of International Dark Sky Reserves:

http://www.darksky.org/night-sky-conservation/87-international-dark-sky-reserves

Campaign for Dark Skies.

UK based light pollution advisers in association with the British Astronomical Association. http://www.britastro.org/dark-skies/

Use these maps, which show the light pollution in your part of the world:

United States: http://www.jshine.net/astronomy/dark_sky/
Europe: http://www.astronomyforum.net/astronomy-locations/europe/
Australia. http://www.astronomyforum.net/astronomy-locations/australia/

15.8   The History of Astronomy

There are a number of societies and sources that particularly cover the history of Astronomy. There are also a number of books that offer a historical insight to the hobby as well. Some of these are still in print. Many of these be obtained second-hand at reasonable cost.

Portal to the Heritage of Astronomy.

Also contains a number of links to various archives of historical material. http://www2.astronomicalheritage.net/

American Astronomical Society – Historical Astronomy Division. http://had.aas.org/

Society for the History of Astronomy. UK based astronomy Historian Society. http://www.shastro.org.uk/

The Cambridge Concise History of Astronomy. Michael Hoskin. Cambridge University Press. 1999.

The Sleepwalkers: A History of Man's Changing Vision of the Universe. Koestler, A. Arkan 1989. Originally published 1959.

History of Astronomy: From 1809 to the Present. David Leverington. Springer. 1996.

Coming of Age In The Milky Way. Timothy Ferris, Anchor Books. 1988.

Cosmos. Carl Sagan. A great book to accompany Carl's outstanding television series shown in the early 80's. Macdonald. 1985. A DVD of the original TV series is also available.

For a delightful account of what it was like for an American amateur astronomer growing up with a love of astronomy and developing his art: Starlight Nights: The Adventures of a Star Gazer. Leslie C. Peltier. Macmillan. 1967.

Celestial Objects for Common Telescopes. Two volumes, The Solar System & The Stars. Rev. T. W. Webb. Dover. Originally published in 1859, but was still in print until fairly recently.

The Bedford Catalogue. Admiral William H. Smyth. Willmann-Bell. Probably the first "amateur" astronomers observing handbook. First published in 1844 and still in print.

Teneriffe, An Astronomers Experiment. Charles Piazzi Smyth. Cambridge University Press. A new reprint of Piazzi Smyth's report on his trip to Tenerife making high-altitude observations. Originally published in 1858.

Edwin Hubble, The Discoverer of the Big Bang Universe. Sharov A.S, Nivikov I.G & Kiskin V. Cambridge University Press. 1993.

Chapter 16

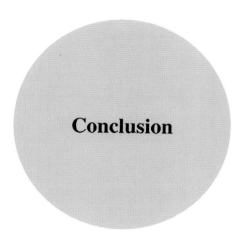

Conclusion

16.1 Keep the Passion Burning

You may not have gone through all the exercises in this book, or read it cover to cover. But the author hopes that he has managed to give you a few pointers along the way and give you as much information and support as possible.

Hopefully the book has inspired you and helped you on your astronomical journey to make further progress and become a much more productive observer? As you are still reading this book you have stuck with it this far. So you probably do have a longer future in our wonderful hobby. Seek advice where necessary, but find your own ways forward to develop. There is no such thing as a stupid question. If you don't know, you don't know. How will you learn if you don't ask?

Almost everyone involved in the hobby is always more than willing to encourage others and answer their questions. If they turn your offers down or snigger at your "ignorance" then they really aren't worth keeping in touch with. Listen to what other more experienced observers have done, but don't always take what they say as "THE FINAL WORD". They really haven't tried everything. However, if you do start to take part in planned observing projects there will be some preferred methodology that should be used for consistency.

Enjoy your hobby in the way that suits you best. Although the author has suggested some ways of working, many of these may not suit your style at all. What works for other observers may just not work for you (and vice versa). Don't gauge your progress by comparing to other observers achievements. It's YOUR hobby, and YOUR development. Not theirs.

D. Eagle, *From Casual Stargazer to Amateur Astronomer: How to Advance to the Next Level*, The Patrick Moore Practical Astronomy Series, DOI 10.1007/978-1-4614-8766-1_16, © Springer Science+Business Media New York 2014

Fig. 16.1 Once you know your quarry, it's easy! (Courtesy of Gita Parekh)

Don't rush or panic! There is so much to observe, you can't see everything very quickly. It takes many, many years to explore the universe properly and you will probably never finish the job. If you miss something special because of the weather, don't be disheartened, there's always something else that will be coming along another time. We are meant to be doing this hobby for enjoyment, to get us away from the pressures of everyday life. So don't put too much pressure on yourself to achieve too much. Just relax and enjoy.

The Moon's brightness will stop you seeing deep-sky objects, meteor showers or comets. It's there whether we like it or not. Learn to love the Moon. Start to observe it. It is the one celestial body that we can see so much intricate detail on its tortured surface, so why not take everything in that it has to offer.

Human light pollution is a growing problem. Yes, we should be fighting it, as it is a battle that is worth winning. Try and learn to work around it, please don't give up because of it. The author lives in a fairly light polluted town, with street lights shining right into his observatory. He still manages to enjoy the hobby to the full. A dose of dark skies at a star party or a week's trip observing always invigorates him and inspires him to carry on while at home.

If you do stick with the hobby, over time you will witness many spectacular things that most Earth inhabitants will be blissfully unaware of. When everything does all come together and you have perfect, clear observing conditions, all your equipment is working perfectly and you are standing out on a dark still night surrounded by the expansive universe in all its glory, what on Earth could be better than that?

16.2   Author's Closing Remarks

Has the information contained within this book helped your development? Do you feel you have developed? In what way? If not, why not? If the failing is due to the advice contained within this book and you have any suggestions for how the book can be improved, please contact the author.

Any mistakes and omissions are purely the author's responsibility and he can only apologize for any that may have made their way into the final print. The author has used many of the sources many times, other useful resources have come to light while researching this book. Despite the care and attention with which the author has applied to his research, mistakes will inevitably set in. You may also, upon reading this book think that something crucial is missing from the information given. Maybe you even disagree with some of the statements made within this book. The author welcomes and even encourages open dialogue and correspondence.

A full list of the URL's included in the book can be downloaded from the authors Web site: www.eagleseye.me.uk/StargazerBookLinks.html

The author also publishes a monthly guide to the night sky "Eagles Eye on The Sky" as a downloadable pdf. This features the coming month's celestial events in a detailed sky diary to aid the observer.

http://www.eagleseye.me.uk/SkyDiary.html

The author also keeps up to date astronomical news and his activities on his astronomy blog:

http://www.eagleseye.me.uk/Blog.html

Expanded versions of the exercises contained within this book are available to download from the authors Web pages, along with a number of other observing guides. More are being added all the time.

http://www.eagleseye.me.uk/Guides.html

Please do get in touch: dave@eagleseye.me.uk

Dave Eagle.   Eagleseye Observatory.   www.eagleseye.me.uk

June 2013.

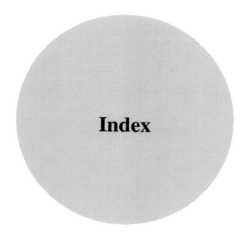

Index

D. Eagle, *From Casual Stargazer to Amateur Astronomer: How to Advance to the Next Level*, The Patrick Moore Practical Astronomy Series, DOI 10.1007/978-1-4614-8766-1, © Springer Science+Business Media New York 2014

Printed in Great Britain
by Amazon

44774073R00152